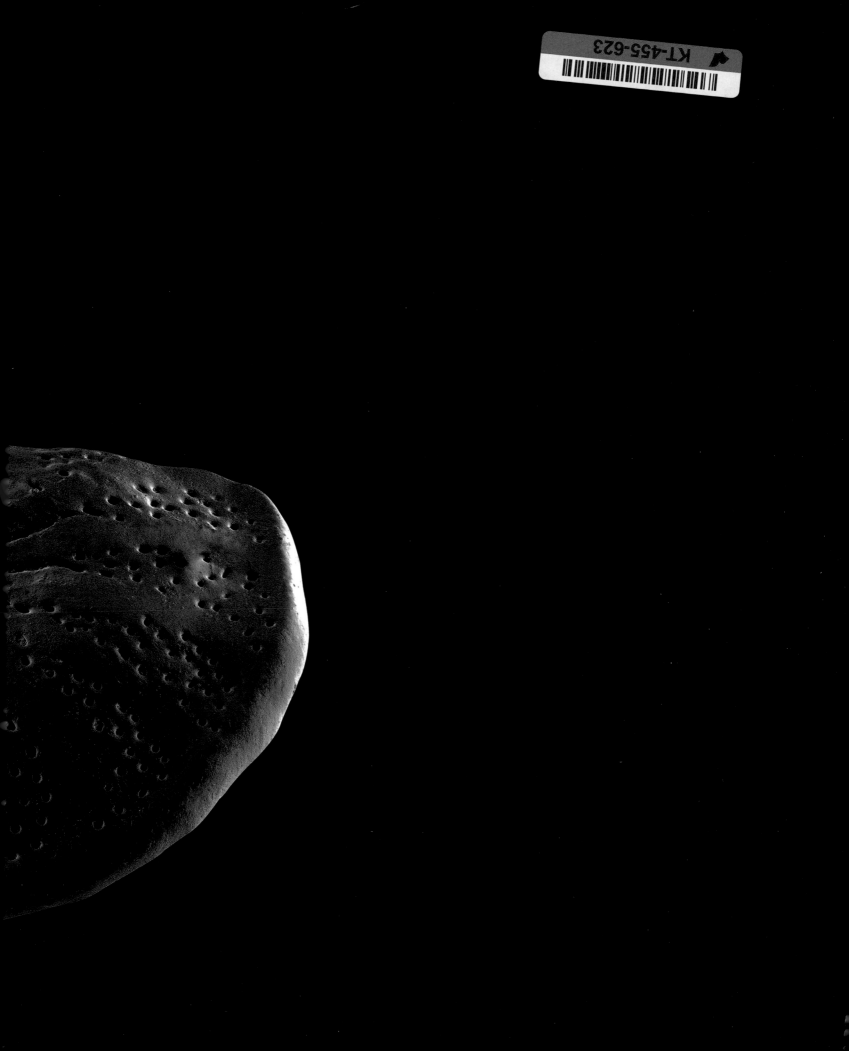

Revealed on these pages is the triumphant power of the scanning electron microscope (SEM) to explore nature and man-made objects at the microscopic level. Combined with the latest techniques microscopists are using to capture the mood and tone of their subjects, and of image artists who use color in order to bring to life the wonder of the microcosmos that surrounds us and is inside us.

Most pictures chosen for this book, unless otherwise described, are taken with SEM's. There is reason for this choice. Other microscopes need the specimen to be sliced thinly, or trapped under glass, before it can be examined, creating a two dimensional picture which appears flat or just a part of the whole. A few examples of these methods are illustrated here such as cell cultures and bird flu virus. But the scanning electron microscope, however, reveals another world entirely, a world familiar to the way we naturally see things, a world with outer surfaces and in three-dimensions. How do SEM's achieve this?

Electron microscopes rely on a beam of electrons, rather than light, to illuminate a subject. And because an electron beam is produced in a vacuum the specimen must be dead and specially prepared. In SEM's, a fine electron beam is scanned across the surface of the specimen, highlighting with precision and detail the complexity of the micro-world: microscopic hairs or pores on leaves, the silicate skeletons that make up tiny plankton, even the intricacies of something as mundane as the weave of water-proof clothing or the coil of a lightbulb filament.

This power of the scanning electron microscope to resolve microscopic structures can be used to explore the beauty of what is too small to see with the naked eye, but it enjoys other applications in biology and medicine, technology and other sciences. For biologists, certain microorganisms may only be properly identified by their filigree shells as seen under SEM, pollen grains by their sculpted surfaces. In medicine, a nerve cell, for example, can be better understood by researchers able to trace the fine paths of nerve fibers in three-dimensions. In engineering, imperfections on a compact disc can be analysed.

Perhaps not realized by the viewer but implicit in these images is that each one has been artificially colored. From a black and white picture produced by SEM, the skilled artist must combine scientific precision with color to make these images come alive. Science has a way of crossing over into art.

Not long ago these artists were hand coloring directly onto black and white SEM prints, or using color gels, or chemical stains. Today, however, in our digital age, technology has introduced powerful computing tools that can not only clean and sharpen an image but with masks and layering give it the subtlety of tones in color and depth of shadow found on these pages.

Neither are all images here attempts at real-life color. The colors may help to highlight important structures, making a complicated picture simpler to understand. Other images may try to mimic real-life, but only as we imagine it — such as in our own bodies. Color is a very subjective concept and may just as well be used to enhance a picture's commercial value. For ease of access depending on how much information the reader requires, every picture has a short title, a longer descriptive one, and an in-depth caption. How many times each image has been magnified under the microscope is represented by an X.

What follows is a journey into the microcosmos as found around us and inside us in its amazing diversity.

Brandon Broll
June 2006

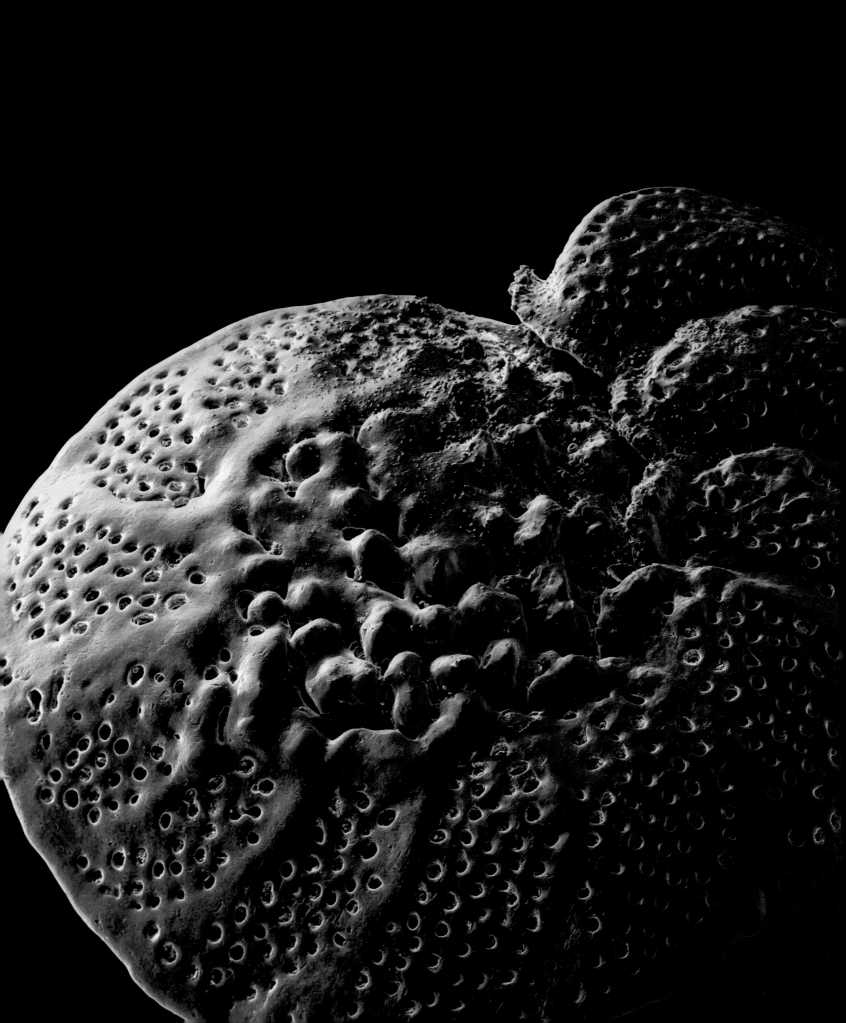

I

Microorganisms

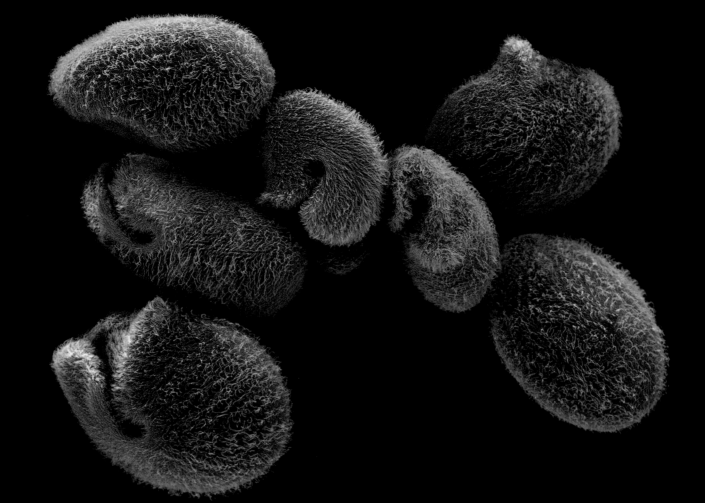

Bresslauides ciliate protozoa

THESE TINY SINGLE-CELLED organisms are found in both water and soil. Their bodies are covered in short microscopic hairs (cilia) used for locomotion. They feed on bacteria and decaying organic matter, helping to clean the soil or water, which they filter through specialized cilia in their primitive mouths known as buccal cavities (seen as a slit). This is a very common species of ciliate often found in large groups.

A GROUP OF CILIATE PROTOZOA (BRESSLAUIDES DISCOIDEUS)

12000✕

Blepharisma ciliate protozoan

COMMONLY VIEWED UNDER MICROSCOPES in science classes, this single-celled organism is found in freshwater. It feeds on bacteria and decaying organic matter, and moves using the tiny cilia on its body. At top is the primitive mouth (buccal cavity) with longer hairs, specialized fused cilia called membranelles, formed into rows which help in feeding. When feeding conditions are good this protozoan can grow to an enormous size and the giant Blepharisma is sometimes seen to feed on its normal sized relatives.

A CILIATE PROTOZOAN (BLEPHARISMA AMERICANUM)

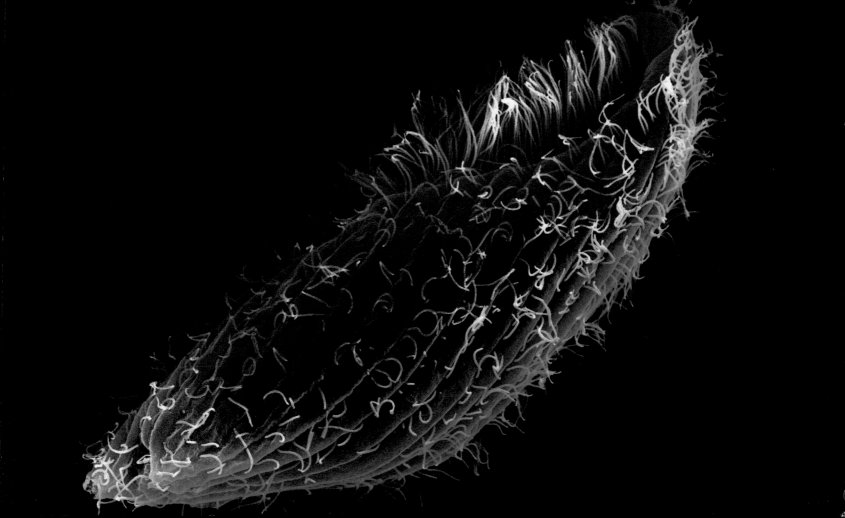

Blepharisma ciliate protozoan

Euplotes protozoan

THIS OVAL SHAPED CILIATE has a transparent body with an open mouth
(at right). It is called a ciliate because of the hair-like cilia that it uses to swim
around its aquatic environment. Different types of cirri, tufts of cilia joined
together, function as a single organelle, and these can be seen on the body and in
the mouth. At upper right is another type of ciliate which this Euplotes may be
attempting to eat. Ciliate protozoa feed on bacteria, other microscopic organisms,
and also scavenge on fragments of food.

EUPLOTES SP. PROTOZOA ARE FOUND IN BOTH SALTWATER AND FRESHWATER ECOSYSTEMS

2000✕

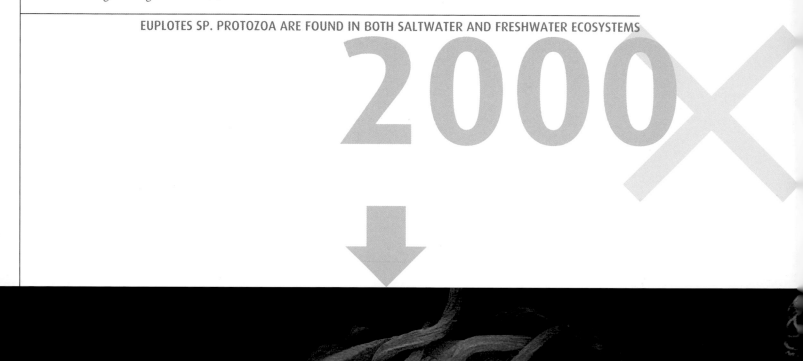

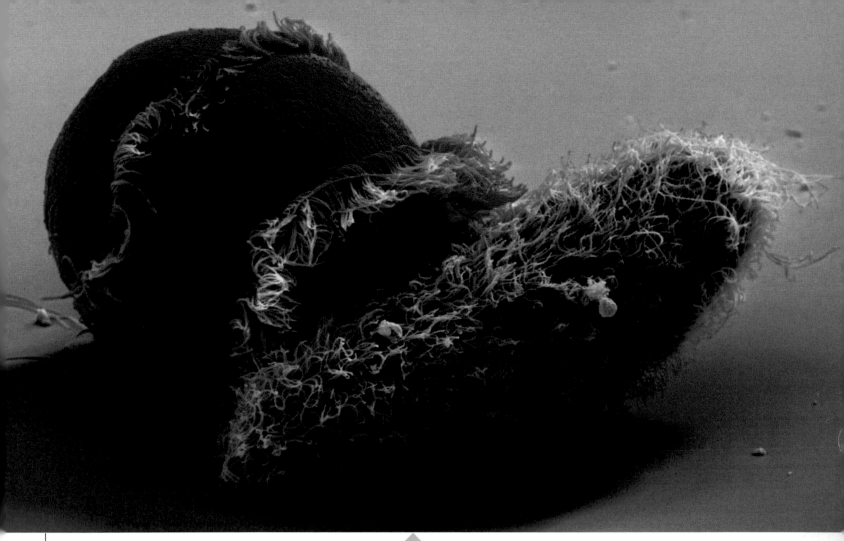

Didinium attacks Paramecium

THESE ANIMALS ARE CILIATE PROTOZOANS, both bearing cilia (microscopic hair-like structures). The barrel-shaped Didinium has two girdles of locomotory cilia, and a prominent snout which is used as a probing and seizing organ during feeding. The Paramecium has cilia across its body used for movement. The Didinium has maneuvered the Paramecium, almost twice its own size, into an easier position for ingestion. The Didinium's snout and gullet are expanding in preparation for the meal.

ONE-CELLED DIDINIUM (BROWN) IN THE PROCESS OF ATTACKING
A ONE-CELLED PARAMECIUM SP. (BLUE)

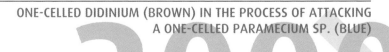

Dendrocometes protozoan

IT LIVES ON THE GILLS OF FRESHWATER FISH and feeds by catching passing particles using its branching tentacles. Protozoa such as this are being used in biotechnology research at Axiva (part of Aventis) in Frankfurt, Germany. They are being studied to see if they can be used to produce useful biologically active chemicals such as enzymes, antibiotics and polyunsaturated fatty acids (PUFAs).

DENDROCOMETES PARADOXUS PROTOZOAN (SINGLE-CELLED ANIMAL)

950✗

Calcareous phytoplankton

THIS SMALL ALGAL ORGANISM (coccolithophore) is surrounded by a skeleton (coccosphere) of calcium carbonate plates (coccoliths). When the organism dies, the plates separate and sink to the ocean floor. Individual plates have been found in vast numbers and can make up the major component of a particular rock, such as the chalk cliffs of England.

EMILIANA HUXLEYI IS FOUND IN SALTWATER ENVIRONMENTS WORLDWIDE

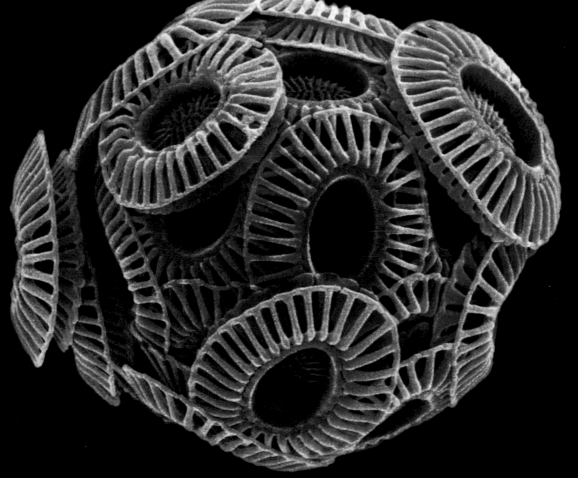

Plankton cell wall

THIS IS A TYPE OF FREE-FLOATING ALGA known as a coccolithophorid, meaning that during its resting phase it produces these plates of calcium carbonate to form a shell around its single cell. It also has a motile phase, when it uses flagella to move. Such algae are found in the sea, forming huge algal blooms when conditions are right.

THE GEOMETRICAL PLATES MAKING UP THE MINERALIZED CELL WALL OF A PLANKTONIC ALGA (CORONOSPHAERA MEDITERRANEA)

4000✕

Radiolarian

RADIOLARIA ARE SINGLE-CELLED protozoans that are found in saltwater plankton. They come in a large variety of intricate shapes, as revealed on the next few pages. This star-shaped one with pointed ray spicules has a spherical skeleton at center through which pseudopodia (false feet, not seen) of protoplasm project. As the animal floats in ocean currents the pseudopodia trap food particles on which the radiolarian feeds.

CENTRAL PART OF THE SHELL OF A STAR-SHAPED RADIOLARIAN

1150✕

Radiolarian

IMPORTANT IN THE SALTWATER plankton food chain, radiolarians float
abundantly in the oceans and can regulate their weight and sink or rise in
the water by having lighter-than-water oil droplets in the protoplasm (not
seen). They build these hard silicate skeletons around themselves. Large
numbers of radiolarian skeletons litter the seabed, becoming fossilized to
form minerals like flint.

DELICATELY SCULPTED SHELL OF A FLASK-SHAPED RADIOLARIAN

960X

Radiolarian

BIOLOGISTS CAN TELL THE EXACT SPECIES of radiolarian by the intricate shape and design of the silicate skeleton, and there are thousands of species living today. Many years ago the great biologist, Ernst Haeckel, made these tiny animals known all over the world in his study of radiolarians, titled: "Art Forms of Nature". It is not hard to see why this saltwater plankton can inspire a scientist to think of art.

SPIKED AND FILIGREE SURFACE OF THE SHELL OF A RADIOLARIAN

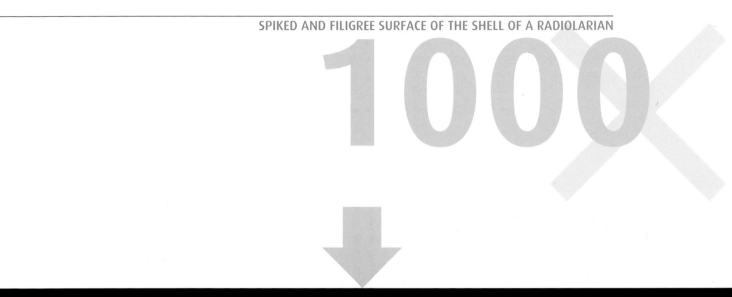

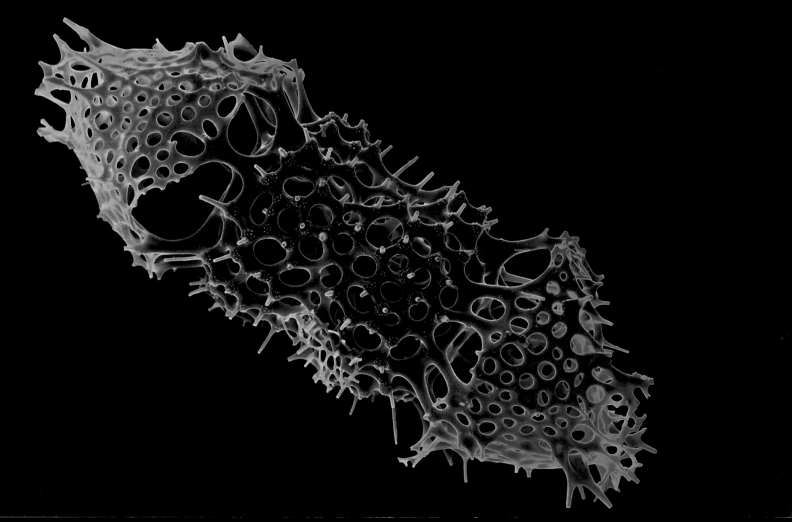

Foraminiferan

THESE SINGLE-CELLED ANIMALS, a relative of the amoeba, are plankton found in saltwater environments and they generally live on the sea floor. They inhabit a calcareous (lime) shell to which new chambers are added as they grow. The shell is penetrated by pores through which the protoplasm of the cell is extruded in long pseudopodia (false feet, not seen). These hair-like feet enable foraminifera to feed on tiny creatures such as diatoms.

FORAMINIFERA CONSTRUCT AND INHABIT SHELLS COMPOSED OF SEVERAL CHAMBERS

400X

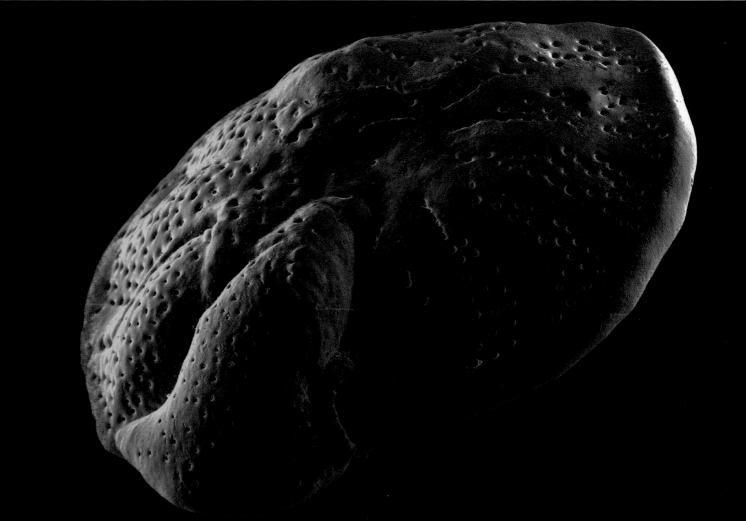

Calcareous phytoplankton

THESE PROTOZOANS CONSISTING of a single-cell are enclosed in a shell with pores through which long pseudopodia (tiny feet, not seen) protrude in order to feed. Found in the sea, these shells eventually form an important component of chalk. In past geological ages, foraminifera were so numerous that their shells, largely composed of calcium carbonate, have formed immense fossil deposits seen today as limestone.

A SPIRAL FORAMINIFERAN SHELL WITH MANY CHAMBERS PENETRATED BY PORES

400✕

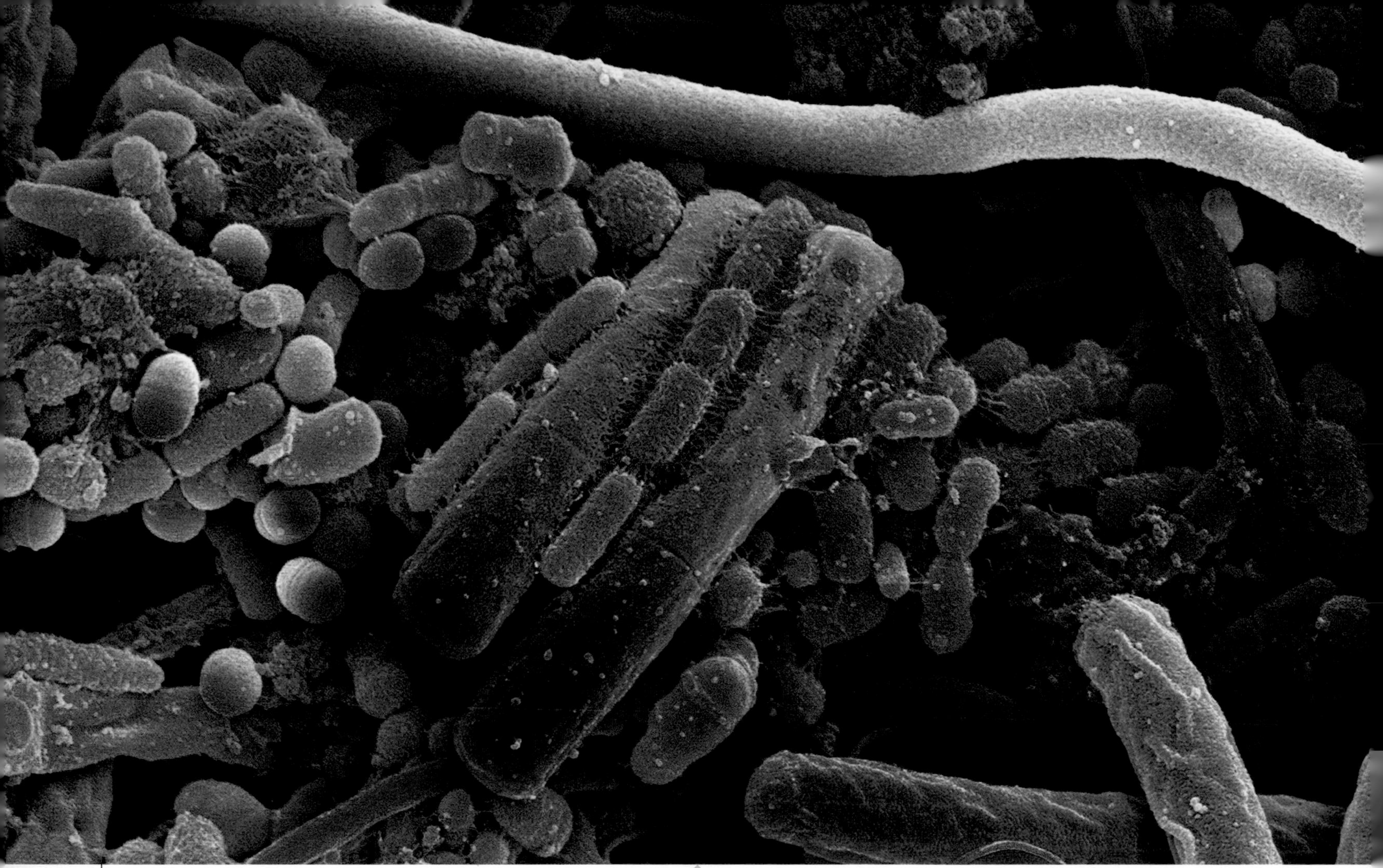

Tongue bacteria

ROUNDED (COCCUS-SHAPED) and elongated (bacillus-shaped) bacteria are seen here with particles of food in the mouth. Large numbers of bacteria can form a visible layer on the surface of the tongue. The mouth contains a large number of bacteria, most of which are harmless or even beneficial. However, some bacteria can cause throat infections or cause the formation of plaque deposits on the teeth, which may lead to decay.

BACTERIA ON THE SURFACE OF A HUMAN TONGUE

23000×

Ulcer bacteria on the stomach

COLONIES OF THIS GRAM-NEGATIVE rod-shaped bacteria are commonly found in the mucosal lining of the stomach in middle-aged people. Its presence has been found in people suffering gastritis (inflammation of the stomach lining), and this bacteria has been linked to stomach ulcer formation. H. pylori may also be a cofactor for gastric cancer as its presence increases the risk of stomach tumors. It can be treated with antibiotic drugs.

HELICOBACTER PYLORI BACTERIA (PINK) ON THE WALL OF THE STOMACH

9960

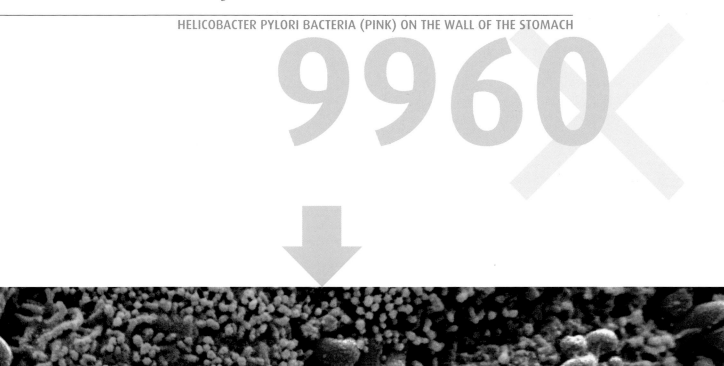

E. coli bacteria

THESE BACTERIA ARE NORMAL INHABITANTS of the human intestine (also animal intestines) and are usually harmless. Under certain conditions E. coli may increase in number and cause infection. Serotypes of E. coli are responsible for gastroenteritis in children, particularly in tropical countries. In adults it is the cause of "traveler's diarrhea" and 80% of all urinary tract infections. It is also the organism most used in genetic studies.

ROD-SHAPED, GRAM-NEGATIVE BACTERIA, ESCHERICHIA COLI, COMMONLY KNOWN AS E. COLI

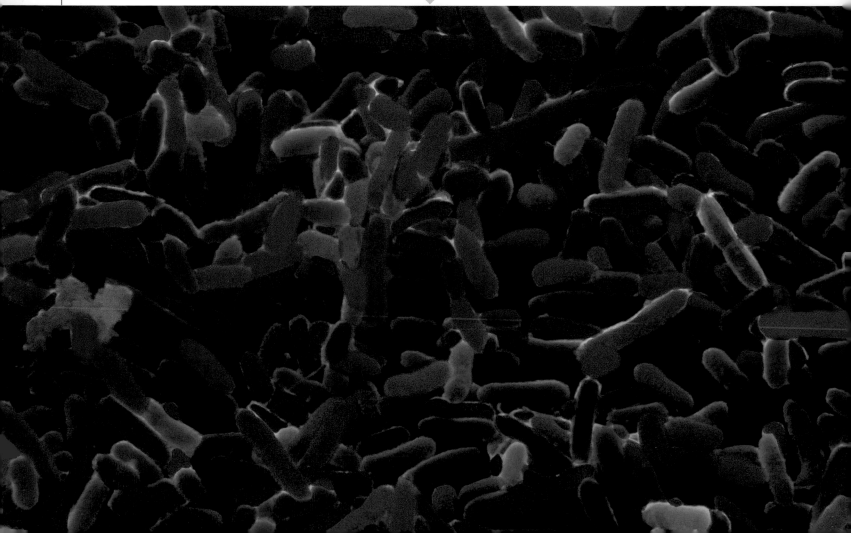

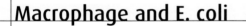

Macrophage and E. coli

MACROPHAGES ARE WHITE BLOOD CELLS which attack many types of disease-causing organisms and clean up cellular debris. They do this by engulfing the organism, such as this bacterium, by a process known as phagocytosis. E. coli bacteria are found in the lower intestines of most mammals. They are usually harmless, but different strains can cause meningitis, septicaemia or food poisoning.

MACROPHAGE WHITE BLOOD CELL (YELLOW) MOVING TOWARD AN
ESCHERICHIA COLI (E. COLI) BACTERIUM (WHITE ROD, LOWER LEFT)

8300✗

Macrophages engulfing E. coli

MACROPHAGES ARE IMPORTANT CELLS of the immune system present throughout the connective tissue of the body and around blood vessels. When an infection threatens, they move toward the site and, as a first line of defence, directly attack the bacteria concerned. When a dangerous strain of E. coli is found, the macrophages engulf these bacterial rods in a process known as phagocytosis and in this way the infection is fought.

WHITE BLOOD CELLS (MACROPHAGES, YELLOW) ENGULFING ESCHERICHIA COLI BACTERIA (RED RODS)

3000×

Macrophage engulfing E. coli

THE TWO BACTERIAL RODS of E. coli are considerably smaller than the attacking macrophage (at top) whose pseudopodia ("false feet" projections from its surface) are moving toward and touching the bacteria. In this way the pseudopodia trap these foreign organisms and engulf them where they are destroyed by enzymes. Macrophages also play an important role in stimulating other cells of the immune system to respond to foreign agents.

CLOSE-UP OF A WHITE BLOOD CELL (MACROPHAGE, YELLOW) ENGULFING
ESCHERICHIA COLI BACTERIA (RED RODS)

19000X

T-bacteriophages on E. coli

T-BACTERIOPHAGES ARE SPECIFIC PARASITES of E. coli bacteria. The virus attaches
itself to the cell wall of the E. coli cell using its tail. The elongated tail is
a contractile sheath which acts like a syringe to squirt the contents of the
head, the DNA genetic material, into the host cell. Viral DNA commandeers the
genetic machinery of the cell, forcing it to reproduce more bacteriophages.

T-BACTERIOPHAGE VIRUSES ATTACKING A BACTERIAL CELL OF ESCHERICHIA COLI. DOZENS OF
BLUE-COLORED VIRUSES ARE SEEN AROUND THE CELL, EACH HAVING A HEAD AND TAIL

69000✕

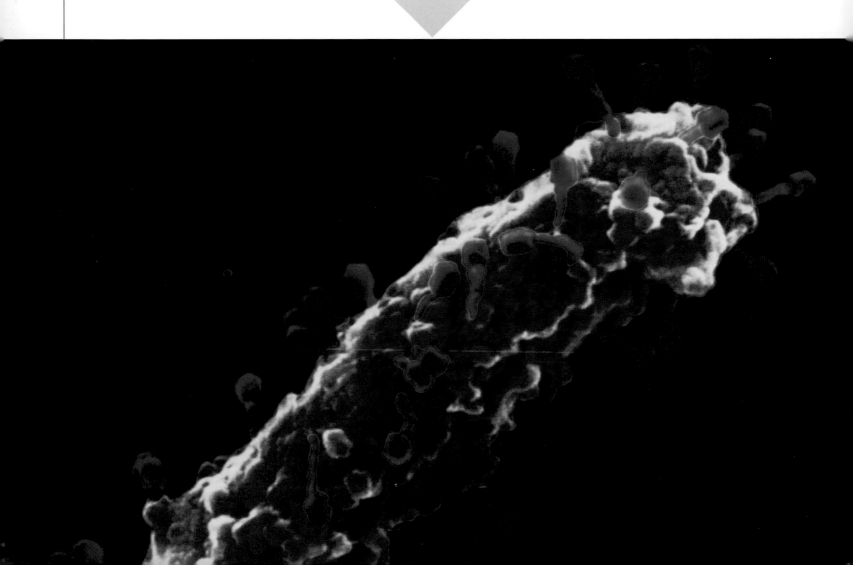

AIDS viruses

THE SURFACE OF THE T-CELL, a type of white blood cell of the immune system, has a lumpy appearance with large irregular surface protrusions. Smaller spherical structures on the cell surface are HIV virus particles budding away from the cell membrane. The virus has infected the T-cell, and instructed the cell to reproduce many more viruses. By this viral budding the T-cell dies. Depletion of the number of T-cells in the blood is the main reason for the destruction of the immune system in AIDS.

T-LYMPHOCYTE BLOOD CELL (GREEN) INFECTED WITH HUMAN IMMUNODEFICIENCY VIRUS (HIV, RED), CAUSATIVE AGENT OF AIDS

20000X

Adenoviruses

ADENOVIRUSES CAUSE SYMPTOMS of the common cold, that is infections of the upper respiratory tract and eyes. The viruses are tiny, about 80 nanometers in diameter, containing a DNA core surrounded by a coat which is a complex icosahedron shape (20 faces). Adenovirus infections are particularly common in children. It is also believed that this virus is associated with the transformation of some types of ordinary cells into cancerous cells. Adenoviruses were first isolated from human adenoidal glands.

GROUP OF ADENOVIRUSES (YELLOW) ON THE SURFACE OF A RED BLOOD CELL OF A CHICKEN

1900

Coronavirus particles

CORONAVIRUSES ARE RESPONSIBLE for causing diseases such as gastroenteritis and SARS (severe acute respiratory syndrome). SARS is a fatal disease that first appeared in China in 2002. The coronaviruses take their name from the crown (corona) of surface proteins which they use to attach to and penetrate their host cells. The virus particles (virions) enter a host cell and use its own cellular machinery to make more copies of the virus. These copies then burst out of the cell, killing it, before infecting other cells.

CORONAVIRUS PARTICLES (YELLOW) ON THE SURFACE OF A CULTURE CELL (BLUE)

42000✕

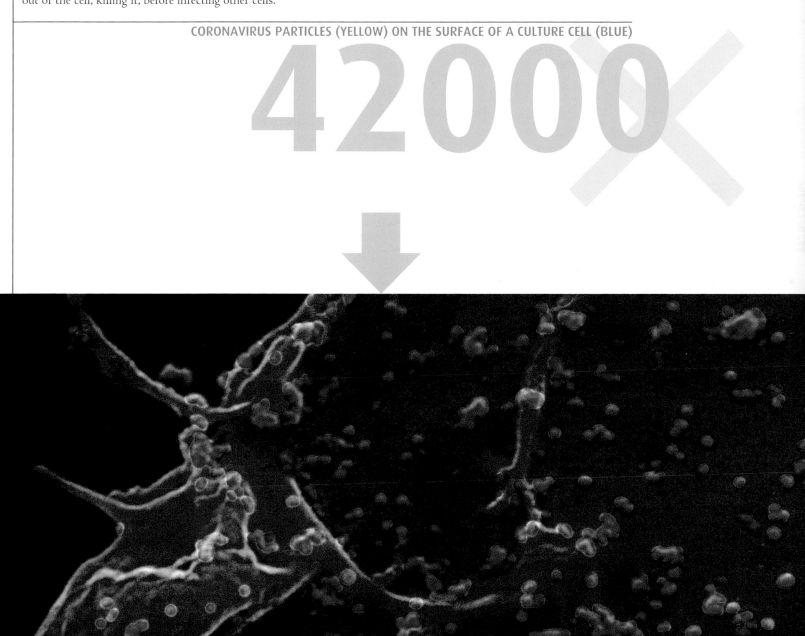

Vaccinia virus particles

THE NUCLEOCAPSIDS (COATS, GREEN) have split and DNA (deoxyribonucleic acid, orange) has been released into the infected cell's cytoplasm. This method of virus reproduction is unusual as most other viruses replicate in the host cell's nucleus. The yellow structures are the cell's golgi bodies, an organelle that the virus uses to make its envelope. Vaccinia belongs to the orthopoxvirus group. It causes cowpox, a disease of cattle and humans, which produces skin lesions. It was first used by Jenner in 1796 to vaccinate against smallpox, a related but more deadly human disease.

TRANSMISSION ELECTRON MICROGRAPH OF VACCINIA VIRUS PARTICLES REPLICATING INSIDE A CELL'S CYTOPLASM

126000✕

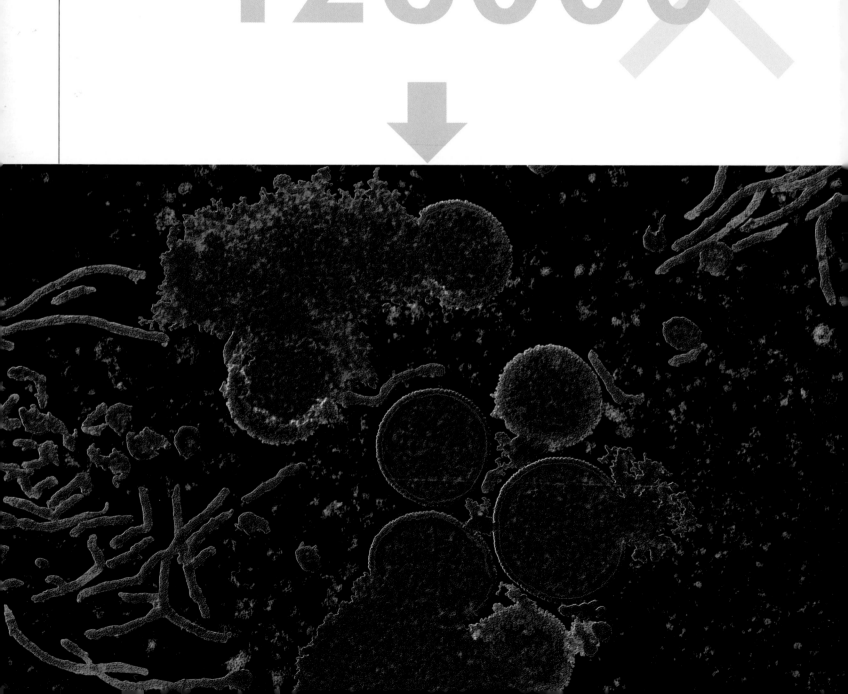

Bird flu virus

THIS LETHAL STRAIN OF BIRD FLU (H5N1: which usually infects poultry) began infecting humans in Hong Kong in 1997. Between 2003 and 2005 it caused the deaths of over 60 people, killing around half of those infected in South-east Asian countries. Humans can catch the virus if they are in direct contact with infected poultry, and migrating wild birds can transport the virus across continents. Fears are that the virus may mutate into a human transmissible form which could lead to millions dying worldwide, such as occurred in the Spanish flu pandemic of 1918. The blue culture cells seen here, Madin-Darby canine kidney cells, are used in vaccine production and viral research.

TRANSMISSION ELECTRON MICROGRAPH OF AVIAN INFLUENZA A VIRUS
PARTICLES (RED) AMONG CULTURE CELLS (BLUE)

66375

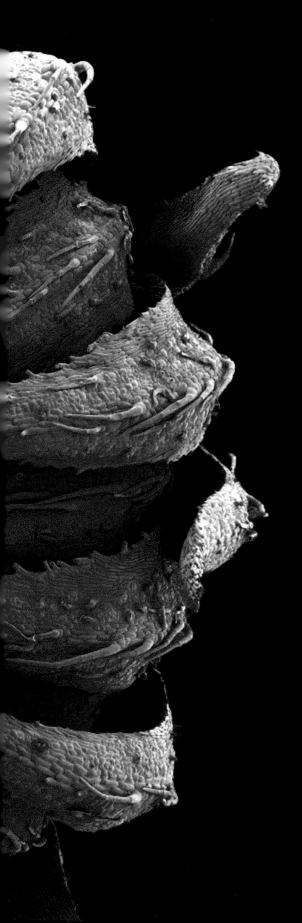

II

Botanics

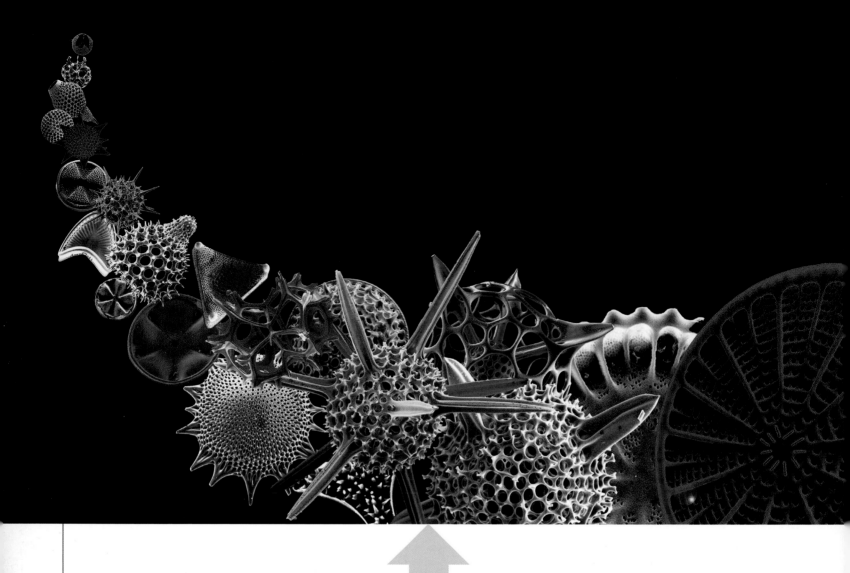

Diatoms

DIATOMS DISPLAY A WIDE VARIETY of shapes and sizes, and as a group they comprise about 10,000 species. As single-celled algae they form an important part of the phytoplankton at the base of the saltwater and freshwater food chains. Known for the fine patterns of the cell wall (frustule) and the tiny holes (striae) within it, this image reveals how different in structure diatoms can be. However, there are just two basic forms: round (centric) and elongated (pennate) diatoms.

AN ASSORTMENT OF DIATOM SPECIES. SOME RADIOLARIA AND FORAMINIFERA ARE ALSO SEEN

500X

Diatom

Diatoms are a group of photosynthetic, single-celled algae found in both saltwater and freshwater ecosystems. The characteristic feature of diatoms is their intricately patterned, glass-like cell wall, or frustule. The frustule consists of two halves which fit together like the lid and bottom of a box, with the "lid" seen here. It often has rows of tiny holes, known as striae. In elongated (pennate) diatoms, as in this one, these striae are arranged on two sides of a central axis.

MINERALIZED CELL WALL OF THE PENNATE DIATOM, MASTOGLOIA SPLENDIDA

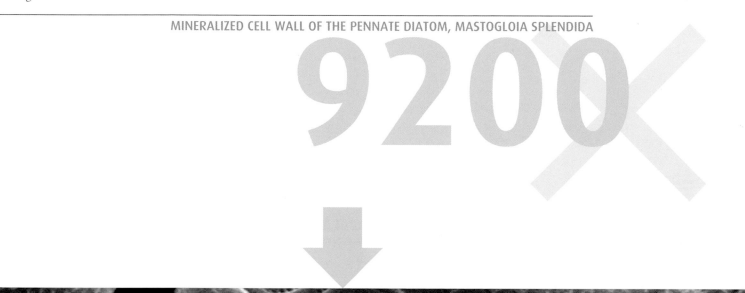

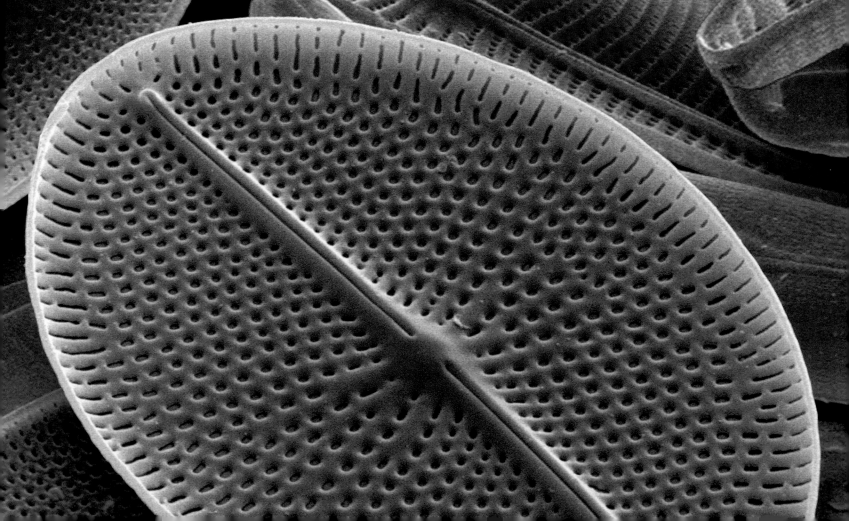

Diatom

PLANKTONIC IN THE WAY THEY FLOAT on water in saltwater and freshwater environments, diatoms are tiny algae each made up of a single cell. Enclosing the cell is an intricate cell wall (frustule) of silica which provides protection and support. The tiny holes (striae) within it give the cell contact with the outside and in this round (centric) diatom the holes are arranged radially from the center outward forming star-shaped patterns.

PATTERNED CELL WALL OF A CENTRIC DIATOM

1760

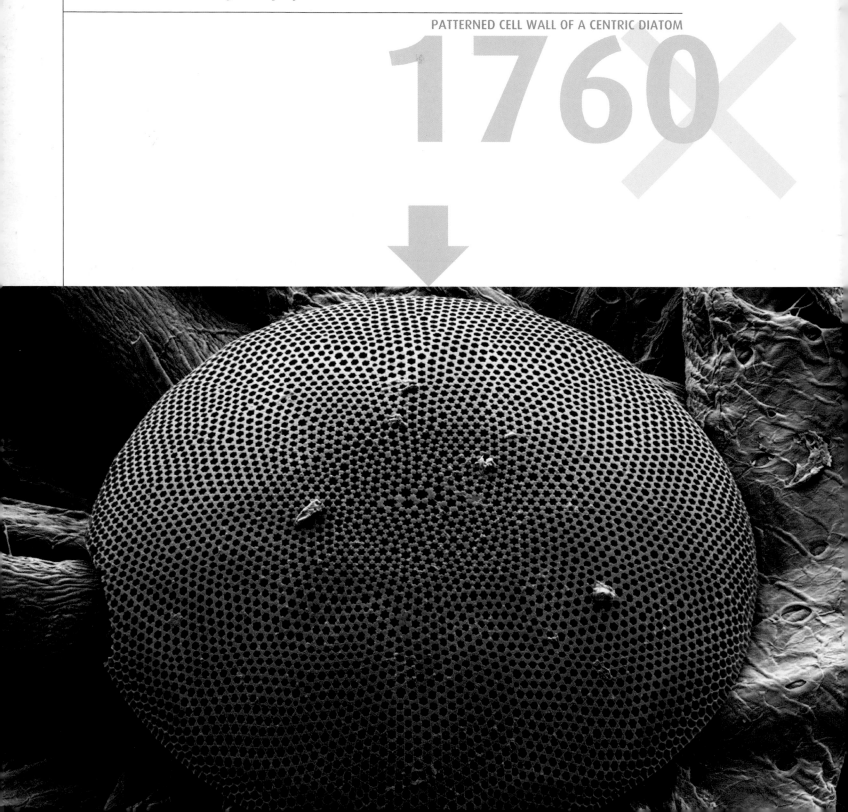

Aspergillus fungus

THE FUNGUS IS MADE UP of fungal threads (hyphae, gray) with conidiophores (fruiting bodies, brown) at the tips. The conidiophores are made up of chains of conidia (spores) which are dispersed on the wind. Inhalation of spores by people with a weakened respiratory system, for example asthmatics or those with cystic fibrosis, leads to an allergic reaction known as aspergillosis. A. fumigatus usually grows on decomposing organic matter.

FRUITING BODIES OF THE MEDICALLY IMPORTANT FUNGUS, ASPERGILLUS FUMIGATUS

12000X

Mushrooms spores

SPORES ARE THE REPRODUCTIVE structures of the fungus, formed by and released from the fruiting bodies, in this case a mushroom. In particular the spores are found in the gills on the underside of the cap of the mushroom. "Agaricus" means gilled mushroom, whereas "bisporus" refers to the fact that each white structure here produces two spores. This is the common cultivated edible mushroom also known as the white button pizza mushroom.

ROUND SPORES (BROWN) OF THE AGARICUS BISPORUS FUNGUS

6900✕

Lichen

LICHENS ARE SYMBIOTIC ORGANISMS formed from both a fungus and an alga that mutually benefit one another. The alga is contained within the hyphal filaments of the fungus and is therefore protected from harsh conditions, in particular from drought and intense light. The alga carries out photosynthesis and in this way supplies the fungus with nutrients. Lichens are slow growing organisms living to immense ages on tree trunks, soil and rocks.

THE ROUND STRUCTURES ARE THE FRUITING BODIES (APOTHECIA) WHICH CONTAIN THE REPRODUCTIVE SPORES

110X

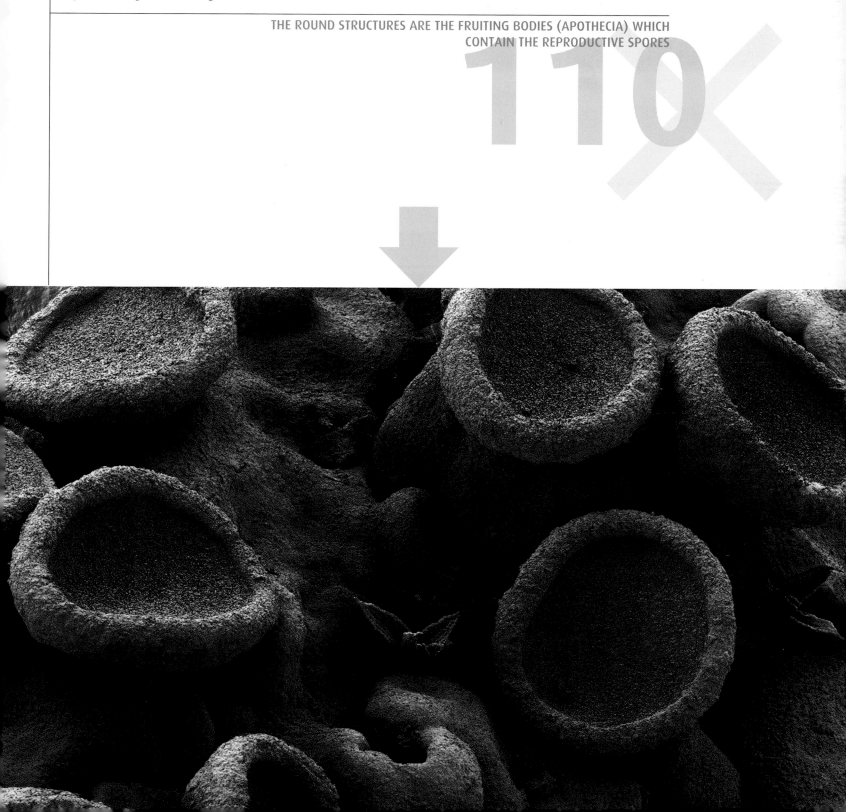

Moss spore capsule

MOSSES REPRODUCE BY MEANS OF SPORES (small pink spheres) at certain times during their life cycle. The spores are dispersed from the mouth of the capsule, being flung away from the plant as the numerous rays (light brown) open. Each capsule is borne on a spiraling stalk which is coiled up in dry weather, only unrolling in damp conditions. Club mosses are common on recently-burned areas and on alkali soils.

PART OF THE OPEN MOUTH OF A CAPSULE (SPORE CASE) OF A CLUB MOSS (FUNARIA HYGROMETRICA)

520✗

Royal fern spores

THE ROYAL, OR FLOWERING, fern produces new fertile fronds in autumn, unlike other ferns which reproduce in spring or summer. On these separate branches are fertile leaves totally covered in spore-producing sporangia. The spores are shaped like wings for dispersal on the wind, and when blown away they last for only two days before dying. The Royal fern is found along streams, in bogs and other wet areas.

THREE SPORES OF THE ROYAL FERN (OSMUNDA REGALIS)

12000×

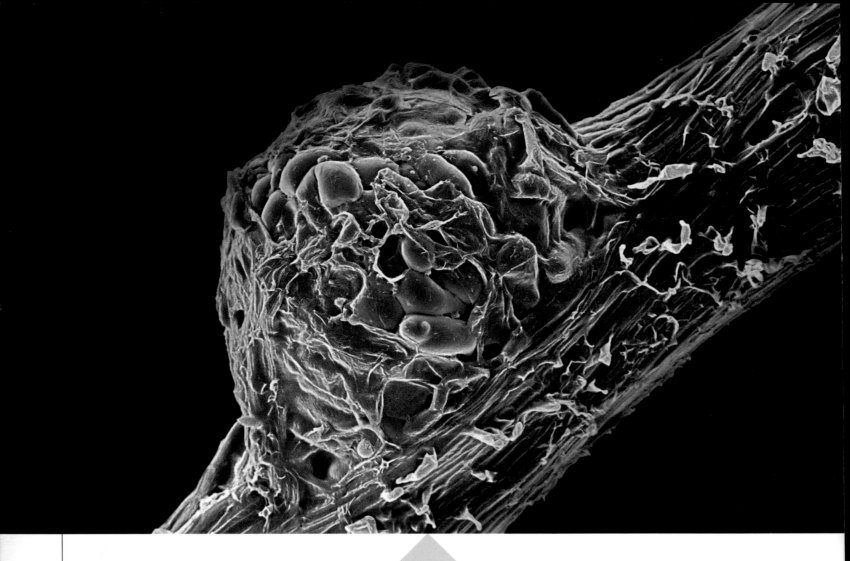

Root nodule

THE PLANT AND THE BACTERIA have a symbiotic relationship. The bacteria converts (fixes) atmospheric nitrogen in the soil to a usable organic form which the plant can utilize. The plant cannot carry out this process itself, but it is vital for the production of amino acids, the building blocks of proteins. In return the plant passes carbohydrates produced during photosynthesis to the bacteria for use as an energy source. The bacteria enters the plant through its root hairs, where an infection thread leads it to the nodule.

ROOT NODULE ON A PEA PLANT (PISUM SATIVUM) CAUSED BY THE NITROGEN-FIXING
SOIL BACTERIA RHIZOBIUM LEGUMINOSARUM

280X

Root of Iris

THE PINK CORTEX SURROUNDS a central stele of vascular tissue defined by a white ring of cells known as the endodermis. Within this, large metaxylem vessels (empty cells) and smaller protoxylem vessels (white) carry water and minerals absorbed by the root to the rest of the plant. The groups of phloem cells (dark purple) carry sugar in the reverse direction, from leaves to root. This vascular arrangement of cells is typical of monocotyledonous plants. Iris germanica is known as the purple flag plant or London flag.

POLARIZED LIGHT MICROGRAPH OF A CROSS-SECTION THROUGH THE ROOT OF IRIS GERMANICA SHOWING THE CENTRAL VASCULAR TISSUES

165

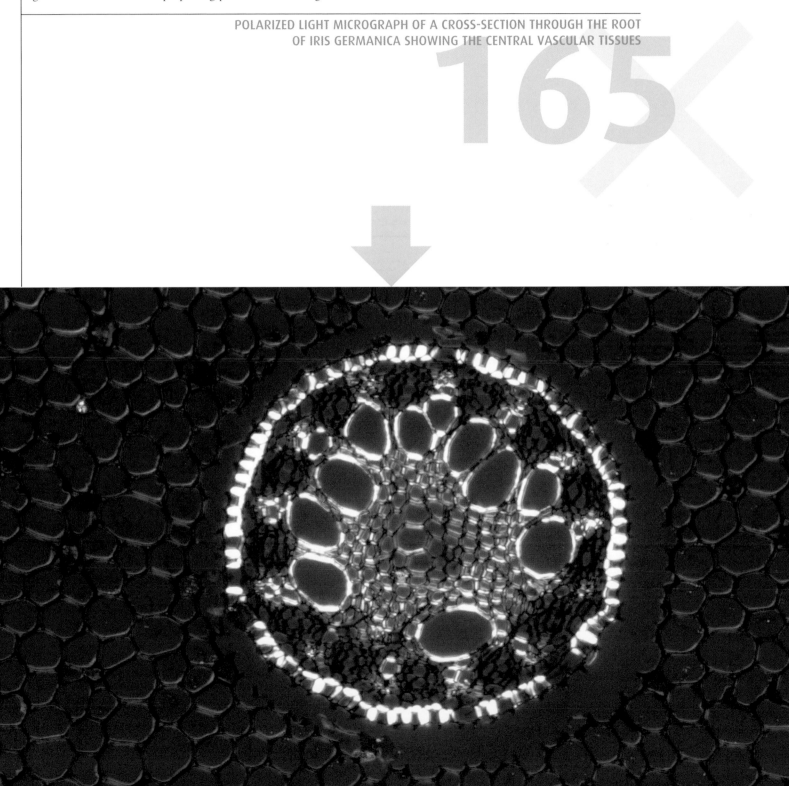

Stem of Dead Nettle

THE HOLLOW CENTER IS PRODUCED by the collapse of thin-walled pith cells.
Surrounding this core is a layer of unspecialized cortical cells – the cortical
parenchyma (yellow & blue). Vascular bundles (red) serve to conduct water
and nutrients: one near each of the 4 corners, and one halfway along each side of
the stem. In the extreme corners of the stem, the best position mechanically, small
collenchyma cells (green) are visible. The hairs on the outside of the stem are
designed to discourage insects from climbing the plant.

CUT STEM OF THE WHITE DEAD NETTLE (LAMIUM ALBUM)

38

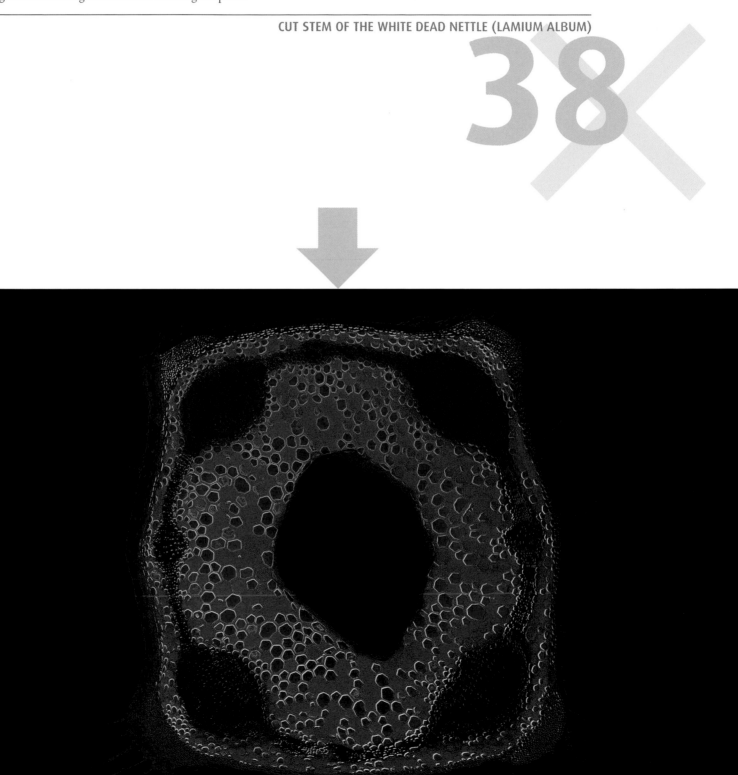

Cactus hairs

THE LONG TRICHOMES (hairs) protect the plant from pest damage. Also seen are smaller interlocking projections from the stem's epidermal cells. These smaller projections cover stomata (pores, at left and lower right) in the stem and help to prevent water loss in drought conditions. Stomata open to allow gas exchange between the plant's interior and the atmosphere. The Bunny Ears cactus is native to arid regions of northern Mexico, but is now grown worldwide as an ornamental plant. It also possesses spines (not seen) to deter large herbivores from eating it.

STEM SURFACE OF A BUNNY EARS CACTUS (OPUNTIA MICRODASYS)

575×

Ivy stem

THE STARSHAPED TRICHOMES, a type of modified hair cell, may function in preventing water loss, or in defending the plant against insects. Several stomata are visible, for example in a line from lower left to center. These pores allow for gaseous exchange between the air and the plant's cells. There are fewer stomata on plant stems than on the leaves.

SURFACE OF AN IVY PLANT STEM

100X

Dicotyledon plant stem

THE CENTER OF THE STEM is filled with large xylem vessels for transporting water and mineral nutrients from the roots to the main body of the plant. Five bundles of phloem tissue (pale green) serve to distribute carbohydrate and plant hormones around the plant. The xylem and phloem are surrounded by a ring of parenchyma cells (purple). The stem is enclosed by a layer of epidermal cells, some of which are specialized as hair cells or oil glands.

CROSS-SECTION THROUGH THE STEM OF A GERANIUM PLANT (PELARGONIUM SP.)

30

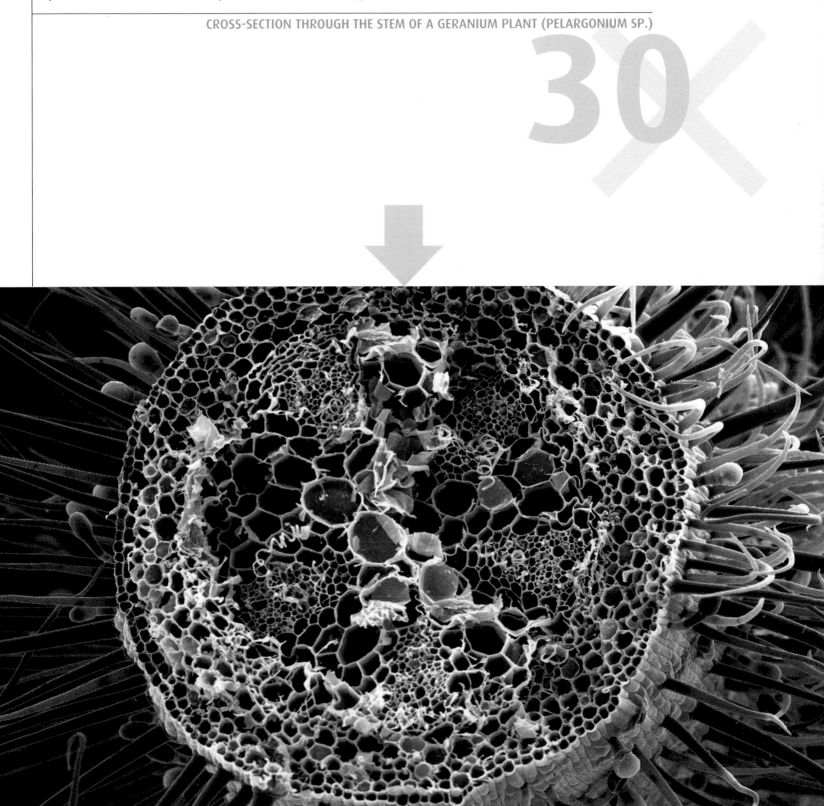

Wood structure

CONCENTRIC RINGS RADIATE OUT from the center (top right) toward the outer
edge of the twig (bottom left). The rings form during successive periods
of growth. The holes are where tube-like structures run along the length of
the twig. These tend to be for structural support near the outer edge of the
twig, and for sap and water transport (in the larger tubes) in the center.

CROSS-SECTION THROUGH A TWIG FROM A HARDWOOD TREE

2K

Clematis stem

TYPICAL OF THIS DICOTYLEDON stem is the ring of vascular bundles known as
the stele, of which five petal-shaped bundles are seen. The large hollow cells
(dark holes) are xylem vessels which transport water from the roots, with
the pith of the stem at bottom center. Secondary thickening has occurred
forming bands of semi-circular sclerenchyma (supporting) tissue, while the
phloem (green) conducts nutrients from the leaves. The white cortex (at
top) is bordered by an outer epidermis (dark brown).

CROSS-SECTION THROUGH A STEM OF CLEMATIS, A CLIMBING GARDEN PLANT

155×

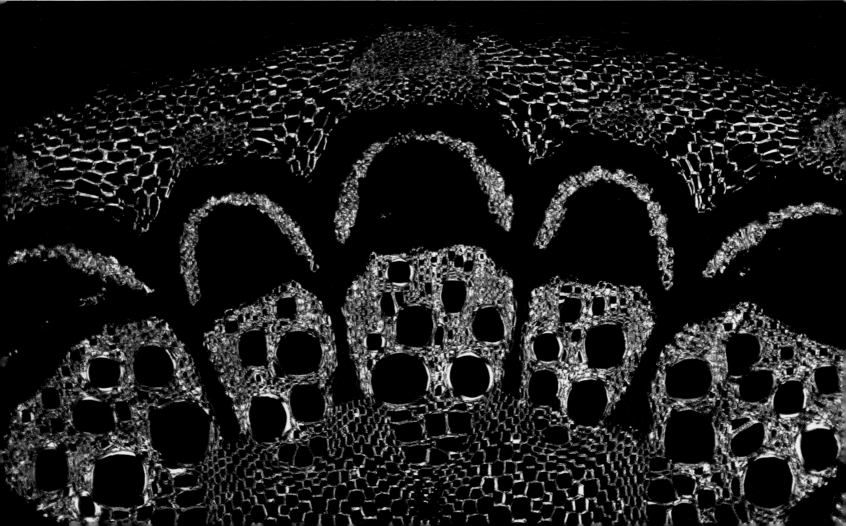

Pea plant xylem

A LONG AND HOLLOW XYLEM vessel runs across the image from left to right made up of individual cells that are dead. The cross walls separating individual cells have broken down to form a continuous tube-like structure, which is capable of transporting water and minerals from the roots to the rest of the plant. These vessels have a unique secondary wall thickening made of lignin which can be, for example, spiral (seen here) running along the outside of the xylem vessel like the coils of a spring.

LONGITUDINAL SECTION THROUGH THE STEM OF A PEA PLANT (PISUM SATIVUM)

2100✕

English Oak heartwood

A TRACHEARY VESSEL (center) has become blocked by papery bladder-like ingrowths known as tyloses. In young sapwood, the vessels carry water and nutrients from the roots to the leaves, but after a few years they become blocked and no longer carry sap. Large vessels like this occur in younger wood. The tissue surrounding the vessel is wood parenchyma, an unspecialized matrix tissue. Oak is classified as a hardwood.

CROSS-SECTION THROUGH THE HEARTWOOD (SECONDARY XYLEM) OF THE ENGLISH OAK, QUERCUS ROBUR

370×

Christmas rose leaf

On the leaf surface is a waxy cuticle (at top) bordered by a row of rectangular epidermal cells. In the body of the leaf (center) are numerous cells containing chloroplasts (green). These are small organelles which are the site of photosynthesis within the leaf. Photosynthesis is the process by which plants use sunlight to turn carbon dioxide into sugars.

SECTION THROUGH THE TOP OF THE LEAF OF A CHRISTMAS ROSE (HELLEBORUS NIGER)

1350✕

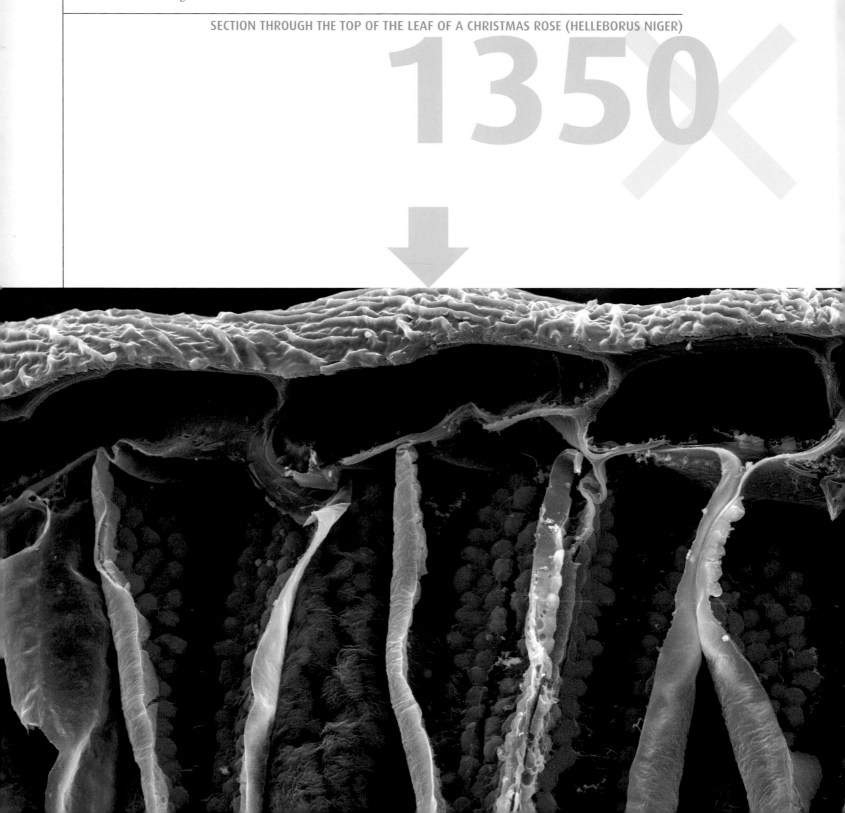

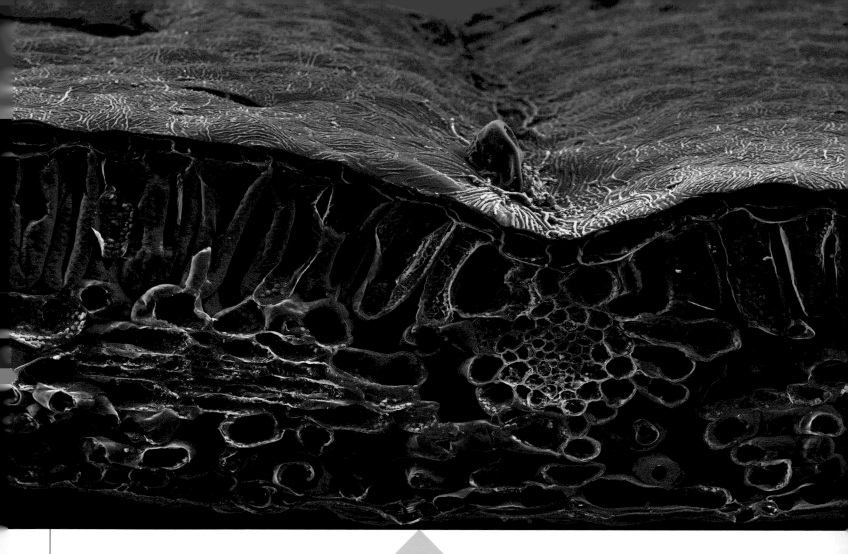

Christmas rose leaf

LEAVES ARE A HIGHLY SPECIALIZED STRUCTURE to trap sunshine and utilize its energy. Beneath the upper epidermal layer of cells with its protective waxy cuticle are two layers which make up the leaf interior: palisade mesophyll (dark green) and spongy mesophyll (light green). Tightly packed palisade cells contain chloroplasts, the sites of photosynthesis. Spongy mesophyll cells have fewer chloroplasts and they are more involved with gas exchange. At lower right (blue) is a vascular bundle, part of the network of veins which transports water and nutrients to and from the leaf.

SECTION THROUGH THE DEPTH OF THE LEAF OF A CHRISTMAS ROSE (HELLEBORUS NIGER)

420X

Nasturtium leaf

NUMEROUS HAIRS (TRICHOMES) cover the surface to protect the plant from small herbivores and reduce the amount of evaporation from the leaf's surface. The branching veins are also seen. The veins consist of the phloem, which transports sugar made in the leaves during photosynthesis to the rest of the plant, and the xylem, which transports water and minerals from the roots to the leaves.

UNDERSIDE OF A NASTURTIUM LEAF (TROPAEOLUM SP.)

45×

Tobacco

THESE LEAF PORES, which are predominantly found on the underside surface of leaves, are the sites of gaseous exchange where carbon dioxide enters the plant and oxygen exits. Tobacco leaves are cured, fermented and aged before being packaged into cigarettes. The plant contains the addictive stimulant drug nicotine.

TOBACCO FROM A CIGARETTE. AT LOWER CENTER AND LOWER RIGHT ARE TWO OVAL STOMATA (PORES)

1265✕

Daisy

PROTECTIVE BRACTS (green) enclose the growing flowerhead of the daisy which is yet to open. Each daisy flower is, in fact, made up of clusters of many tiny flowers called florets.

BUD OF A DAISY FLOWER (FAMILY: ASTERACEAE)

23×

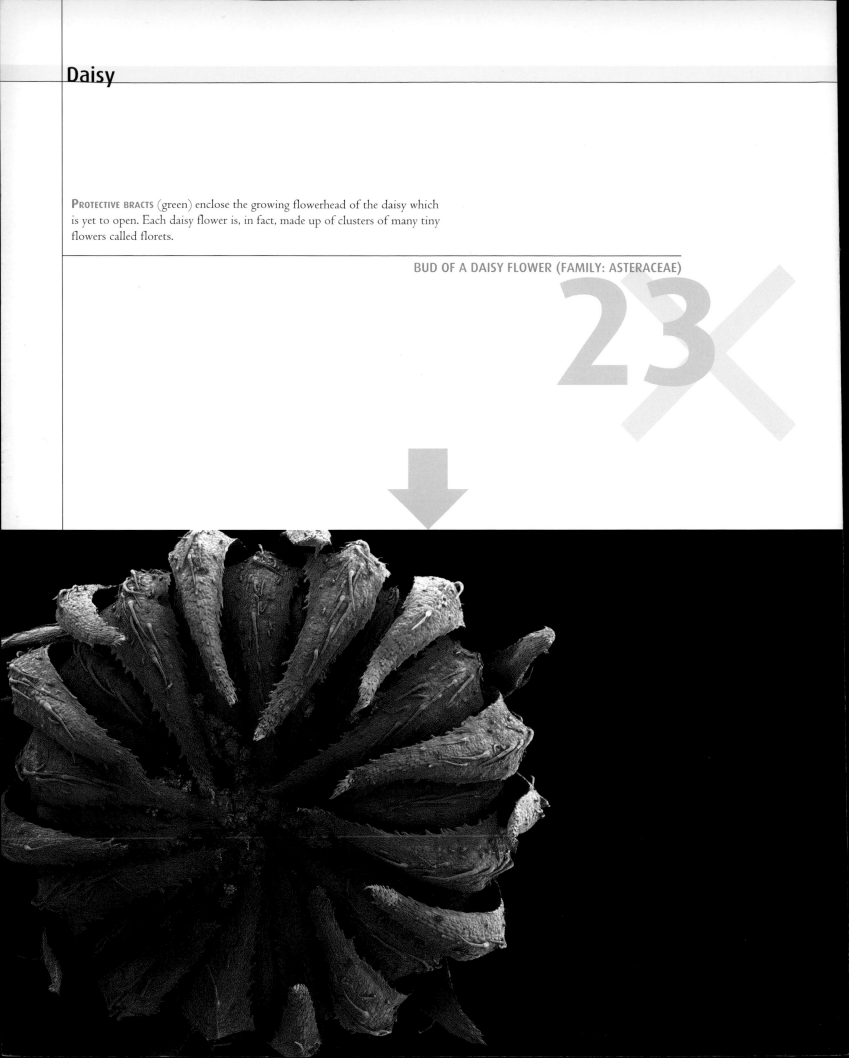

Eggs of an Orchid Cactus flower

WITHIN THE CAVITY OF THE OVARY of this plant 15-100 ovules (eggs) are found, each attached to a stalk called a funicle (green). The eggs (yellow) consist of an embryo sac containing an ovum (egg cell), the female gamete. This is surrounded by tissue known as the nucellus, which provides nutrition for the egg. If fertilized by pollen (the male gamete), each egg can become a seed.

OVARY OF AN ORCHID CACTUS FLOWER (EPIPHYLLUM SP.) SHOWING THE OVULES

175×

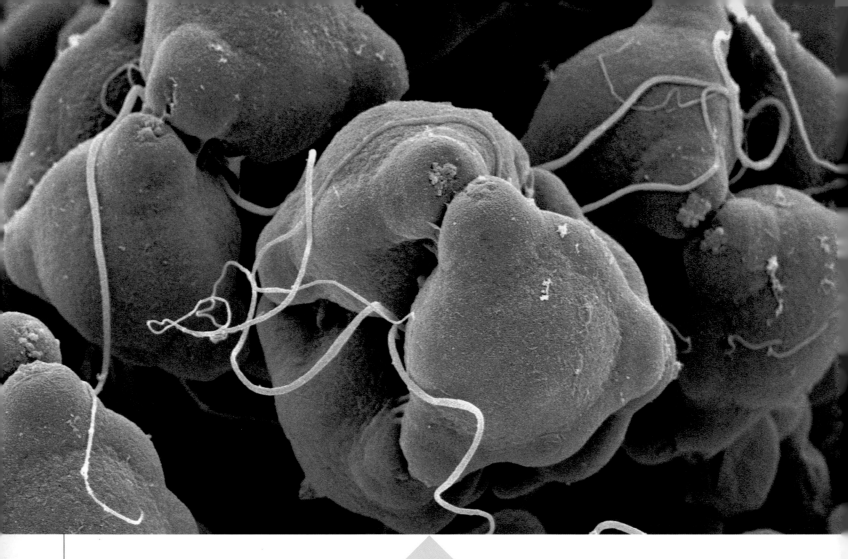

Willowherb pollen

THE TRIANGULAR WIND-DISPERSED grains of pollen are arranged in clusters of four but they separate when carried away on the wind. The shape of each grain assists with its buoyancy in the air. Pollen grains contain the male gametes of the plant and if one lands on the female part of the flower, it fertilizes its ovule. The fertilized ovule grows to become a seed. Rosebay willowherb is a perennial weed common in woods and wastelands.

POLLEN GRAINS OF THE ROSEBAY WILLOWHERB (EPILOBIUM ANGUSTIFOLIUM)

1380✕

Hyssop pollen grains

ANTHERS OF THE FLOWER produce the pollen grains. In this plant the pollen is rounded in shape with six grooves. It is dispersed by insects which carry it to other plants, resulting in the fertilization of the female eggs present in the flower's carpel. Hyssop is a medicinal perennial found growing wild in southern and eastern Europe. Its flowers are used as a tonic and a sedative, while extracts of the leaves have been found to inhibit the replication of the HIV virus.

POLLEN GRAINS OF HYSSOP (HYSSOPUS OFFICINALIS) ON A PETAL

4400✕

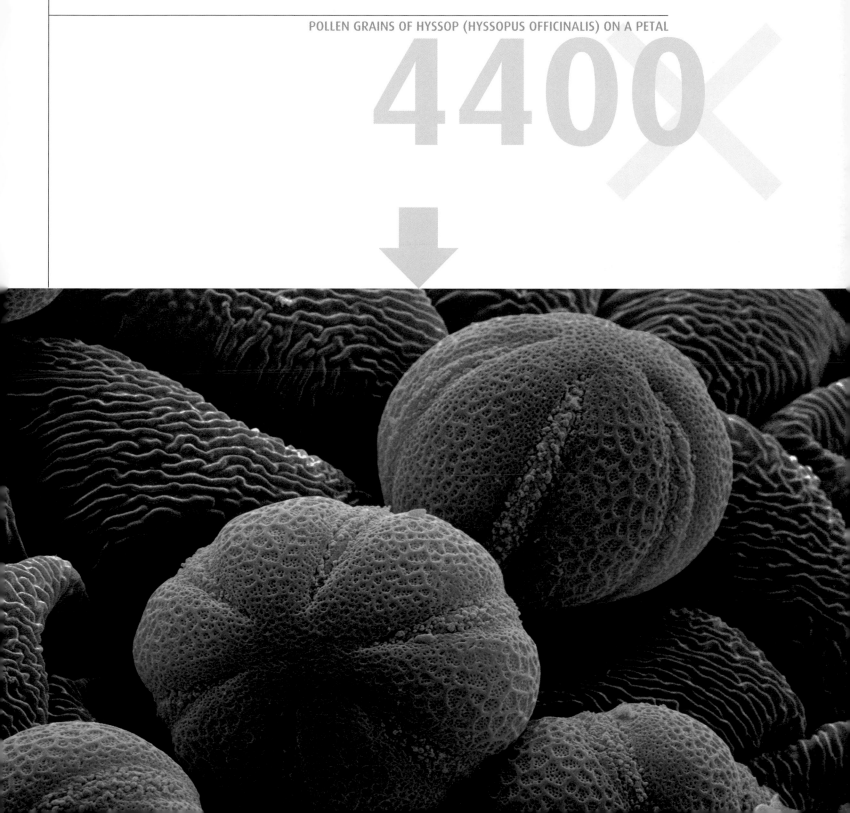

Pollen on a bee's leg

AS THE MALE SEX CELLS of a flowering plant, these pollen grains have attached to the claws and hairs on this bee's leg. Pollen dispersed by insects is often spiky in appearance, designed to adhere to their body when the insect alights on flowers to feed on nectar. As the insect travels from flower to flower, some of the pollen grains are transferred to the female parts of another flower (pollination). Bees are major pollinators of flowering plants.

POLLEN GRAINS (GREEN) ON THE TIP OF THE FOOT OF A BEE

260

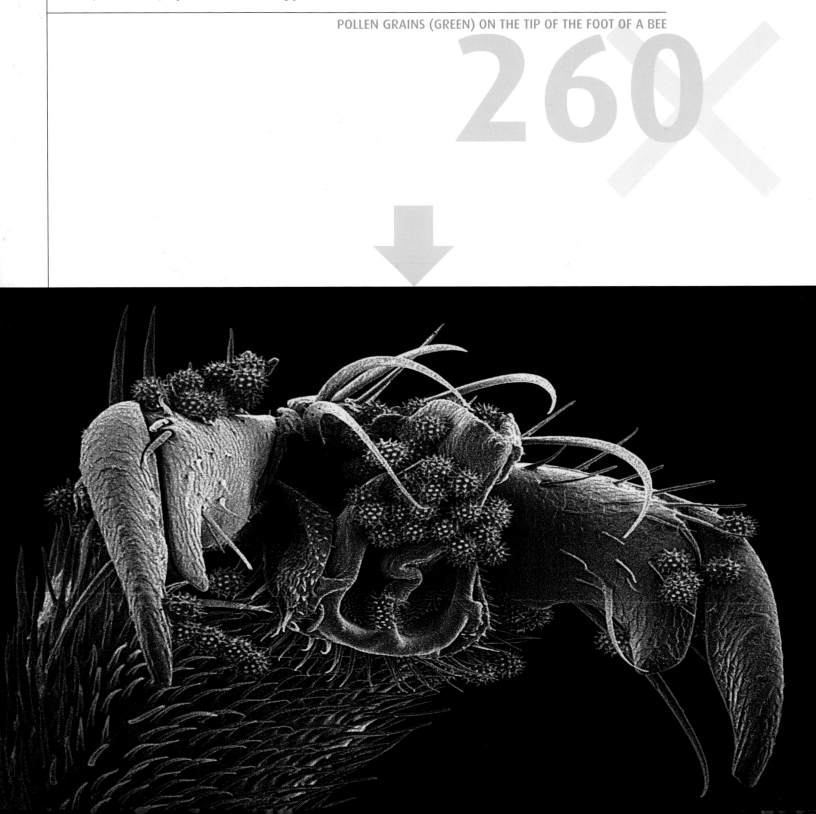

Sunflower pollination

THE STIGMA, PART OF THE FLOWER'S female reproductive structure, is curled over here, with pollen grains (spiky orange balls) adhering to the yellow trichomes (hairs) on its underside. The pollen grains carry the male reproductive cells, designed to travel down the style to the ovary (neither seen) and so fertilize the plant. As a member of the Compositae family, the sunflower's flowerhead is made up of many small florets (true flowers).

POLLEN ON A STIGMA OF A SUNFLOWER PLANT (HELIANTHUS ANNUUS)

500x

Morning Glory pollen

ROUND POLLEN GRAINS, such as these with spikes, are perfectly adapted to attach to insects and other animals which transport them from flower to flower. Here, the pollen has been placed on the pistil (female reproductive part) of another flower in the act of pollination. As its name implies, a Morning Glory flower opens in the morning, allowing it to be pollinated by butterflies, bees and even hummingbirds.

ORANGE POLLEN GRAINS OF A MORNING GLORY FLOWER (IPOMOEA SP.)

165X

Pollinated flower pistil

THE PISTIL IS THE FEMALE reproductive part of the flower. It can be divided into several carpels with a stigma at their tips, in this case three carpels so three stigmas can be seen. The stigma is where pollen lands during pollination. The pollen grains are male gametes (reproductive cells) that can fertilize a female gamete to produce a seed. This requires the male gamete to travel down the style (the structure that supports the stigma) to the ovary at the base of the pistil.

ROUND POLLEN GRAINS (RED) ON THE STIGMAS (YELLOW) OF A
FIELD PENNYCRESS FLOWER (THLASPI ARVENSE)

105✕

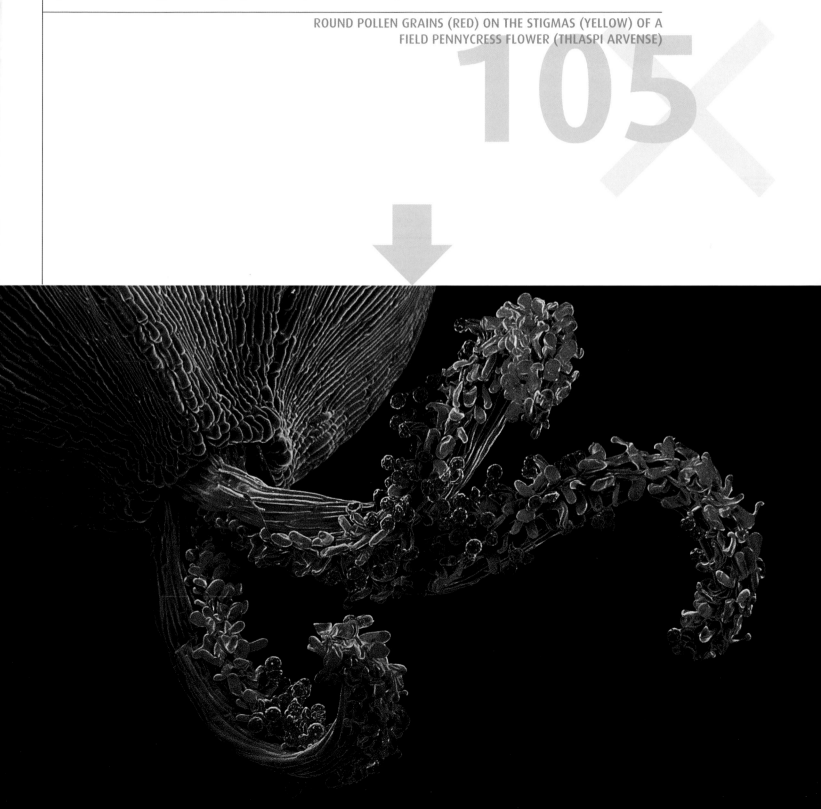

Germinating wheat grain

THE MAJORITY OF THE SEED COMPRISES a store of starch surrounded by a seed coat (yellow). This starch provides the embryo plant with the necessary energy for germination. Germination may be triggered by humidity, temperature or other factors. A root radicle (lower left) emerges from the seed coat and begins to grow down into the soil. The leaf shoot (green) that grows to the surface is surrounded by a protective sheath called the coleoptile.

SEED GRAIN OF GERMINATING WHEAT (TRITICUM SP.)

40X

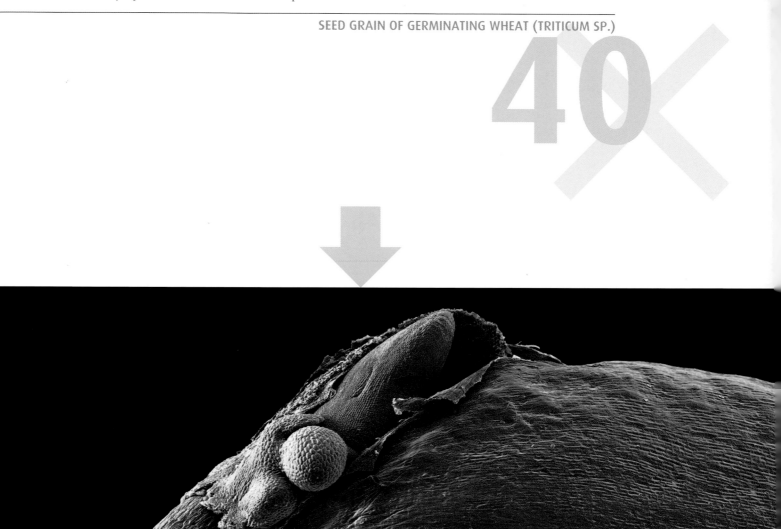

Germinating wheat grain

THE RADICLE IS THE FIRST PART of the seedling to emerge during germination. It emerges through the micropyle (upper center) and grows down into the soil. Roots exhibit positive gravitropism, under the influence of hormones they grow toward the pull of gravity. Germination may be triggered by humidity, temperature or other factors. Root hairs have grown on the radicle. These tiny hairs greatly increase the surface area of the root for uptake of water and nutrients.

ROOT RADICLE OF A GERMINATING GRAIN OF WHEAT (TRITICUM SP.)

115✕

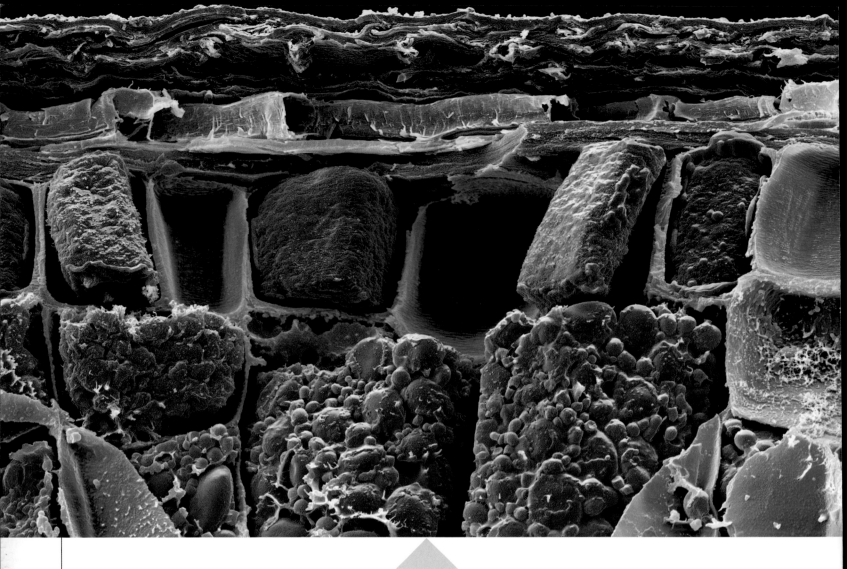

Wheat grain

THE GREATER PART OF THE INTERIOR of the seed is made up of a store of starch grains (yellow) surrounded by cell walls (gray). Above this is a layer of cells containing protein (green). The entire seed is covered in a seed coat (brown). The starch provides the embryo plant with the necessary energy for germination. As an important food crop, wheat grains are usually ground into flour before being used to make bread and other edible products.

SECTION THROUGH A GRAIN (SEED) OF WHEAT (TRITICUM SP.)

1380

Wheat grain

THE IMPORTANCE OF STARCH as a source of energy and nutrients for the seed is obvious here. Starch is stored in the deeper lying cells as starch grains (yellow) surrounded by cell walls (gray). These starch grains (amyloplasts) are synthesized from sucrose, a sugar formed in the leaves during photosynthesis and transported into the seed. Smaller quantities of protein (green) occur in an outer layer of cells, while the seed is covered by a protective coat (brown, at top).

DEEP SECTION THROUGH A GRAIN (SEED) OF WHEAT (TRITICUM SP.)

410X

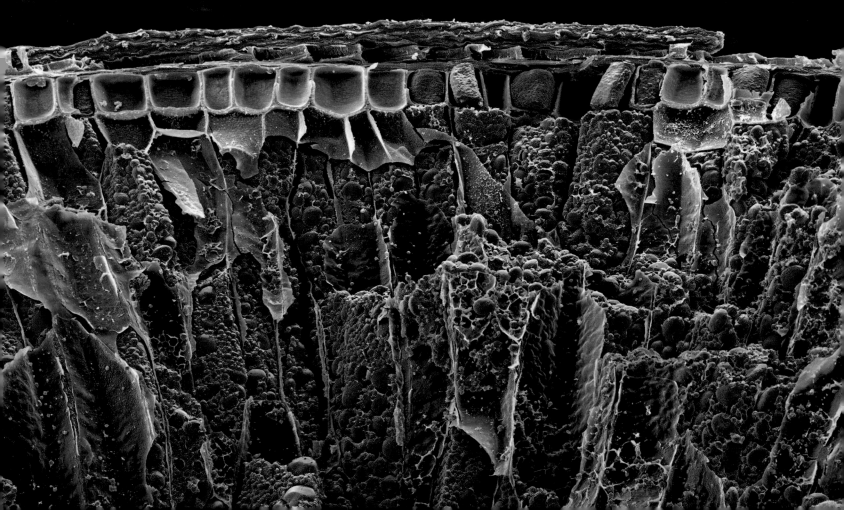

Germinating seed

THE FIRST TWO IMAGES show the emergence of the radicle, the embryonic root, from the seed coat (testa). In the third image, the root grows downward in response to gravity (geotropism), while in the fourth image the embryonic shoot (plumule) grows up against gravity. Its seed leaves (cotyledons, green) will photosynthesize using sunlight. The root "hairs" will help obtain water and nutrients. This is a swede (Brassica napus) seedling.

MAIN STAGES IN THE GERMINATION OF A PLANT SEED. THE SEQUENCE RUNS FROM LEFT TO RIGHT

25✗

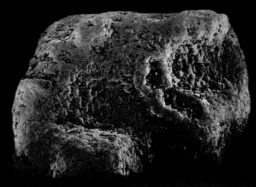

Salt and peppercorn

Salt (sodium chloride) is a crystalline compound commonly obtained from salt mines or by the evaporation of sea water, and used to flavor and preserve food. The peppercorn is the fruit of the tropical vine, Piper nigrum. Ground peppercorns are used as a hot aromatic food seasoning.

SALT GRAIN (BLUE) AND A PEPPERCORN (ORANGE)

19X

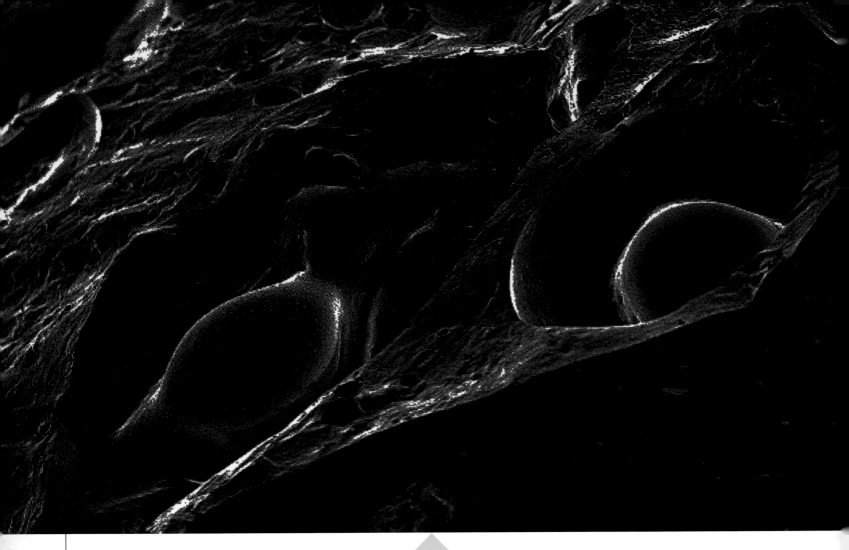

Chocolate ice cream

ALTHOUGH SMOOTH WHEN SEEN by the naked eye, under magnification the surface of an ice cream bar appears moon-like and cratered, due to lipids (fatty acids and glycerol) which form into rounded globules of fat. Saturated fat is common in dairy products such as whole milk and cream. This frozen food is made from a bar of ice cream covered in chocolate.

SURFACE OF A CHOCOLATE ICE-CREAM BAR

390X

Cauliflower

THE COMMON VEGETABLE cauliflower bears large clusters of thick-stalked un-
opened white flowers. In fact multiple flowerheads occur, each with many
individual florets arranged spirally. This spiral shape of the florets describes
a Fibonacci mathematical series, one in which each number is the sum of
the two previous numbers. For example: 0, 1, 1, 2, 3, 5, 8, 13, 21 and so on.
Many natural patterns in flowers can be described using this series.

FLOWERHEAD OF A CAULIFLOWER (BRASSICA OLERACEA BOTRYTIS)

36✕

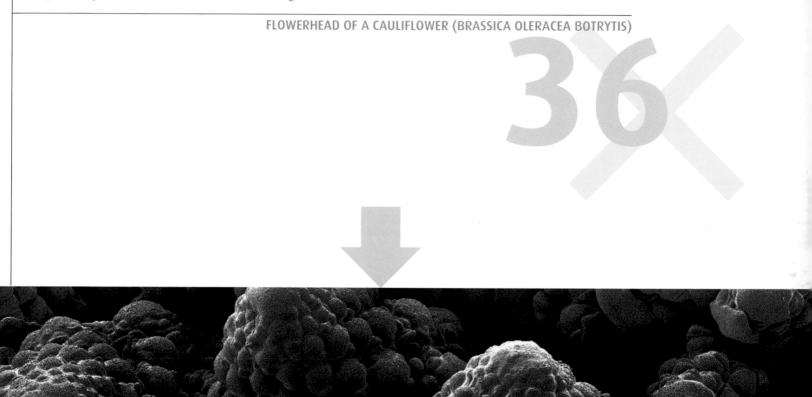

Almond nut

THE FRACTURE PLANE has shown the internal structure, passing through the cells making up the nut. The cells are rounded and each contains lots of fibrous material as well as oil droplets. Almond nuts, which are edible, are also used to produce an oil.

FREEZE-FRACTURED SECTION THROUGH A NUT FROM AN ALMOND TREE (PRUNUS DULCIS)

1650✕

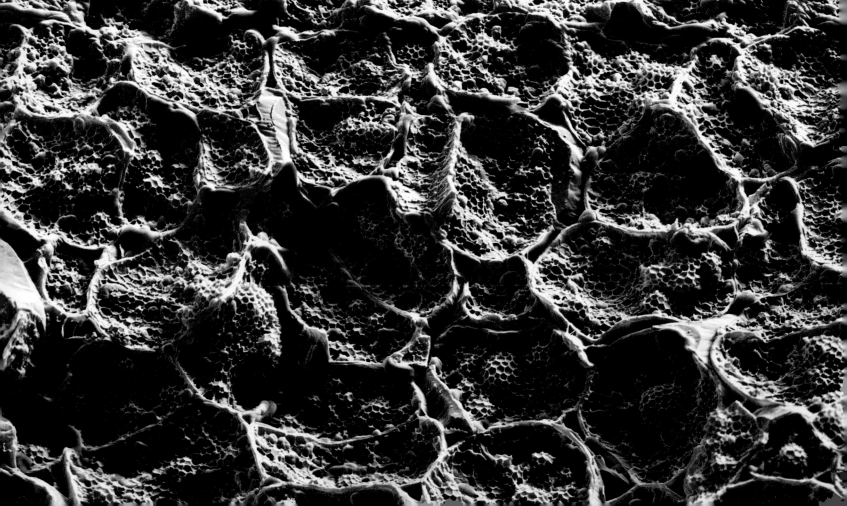

Soy bean

SOY BEANS ARE PRODUCED in seedpods on the soy plant (Glycine max). Starch is stored in spherical kernels (yellow) in the cells and this carbohydrate source provides the germinating seedling with energy. Soy beans are an important nutritional foodstuff comprising between 30-50% protein, 25% carbohydrate (starch) and 15-25% oil from healthy fats low in cholesterol.

FREEZE-FRACTURED SECTION THROUGH A SOY BEAN

1035

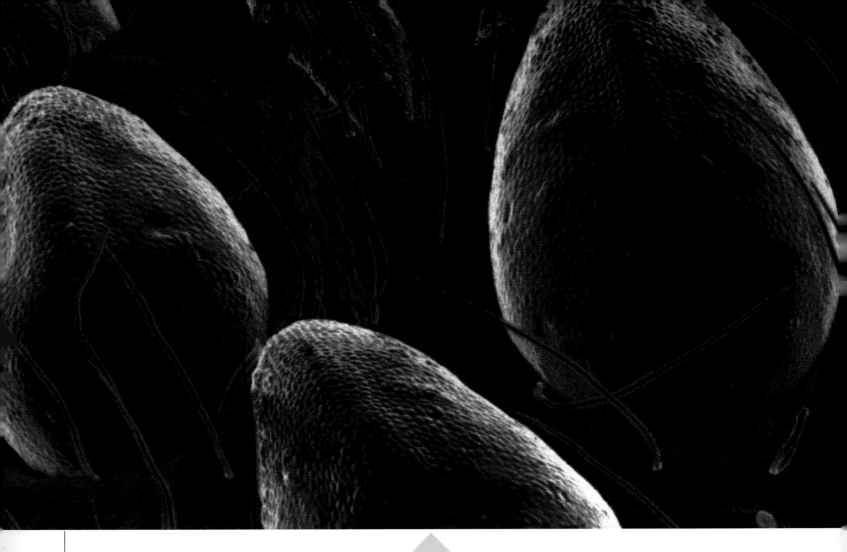

Strawberry

A RIPE RED STRAWBERRY, such as this, is an aggregate fruit, not really a single berry or a single fruit. The true fruits are the achenes on it (seeds, oval shields) while the strawberry itself is actually an enlarged and sweet fleshy receptacle that is edible. As such, strawberries must be picked at full ripeness as they cannot ripen once picked. The brown tubular structures are styles and stigmas of the female reproductive organs protruding between individual achenes. The strawberry plant, pollinated by insects, grows close to the surface of the ground on runners, and its many cultivated varieties are of economic importance.

CLOSE-UP OF THE SURFACE OF A STRAWBERRY FRUIT, FRAGARIA SP.

45✕

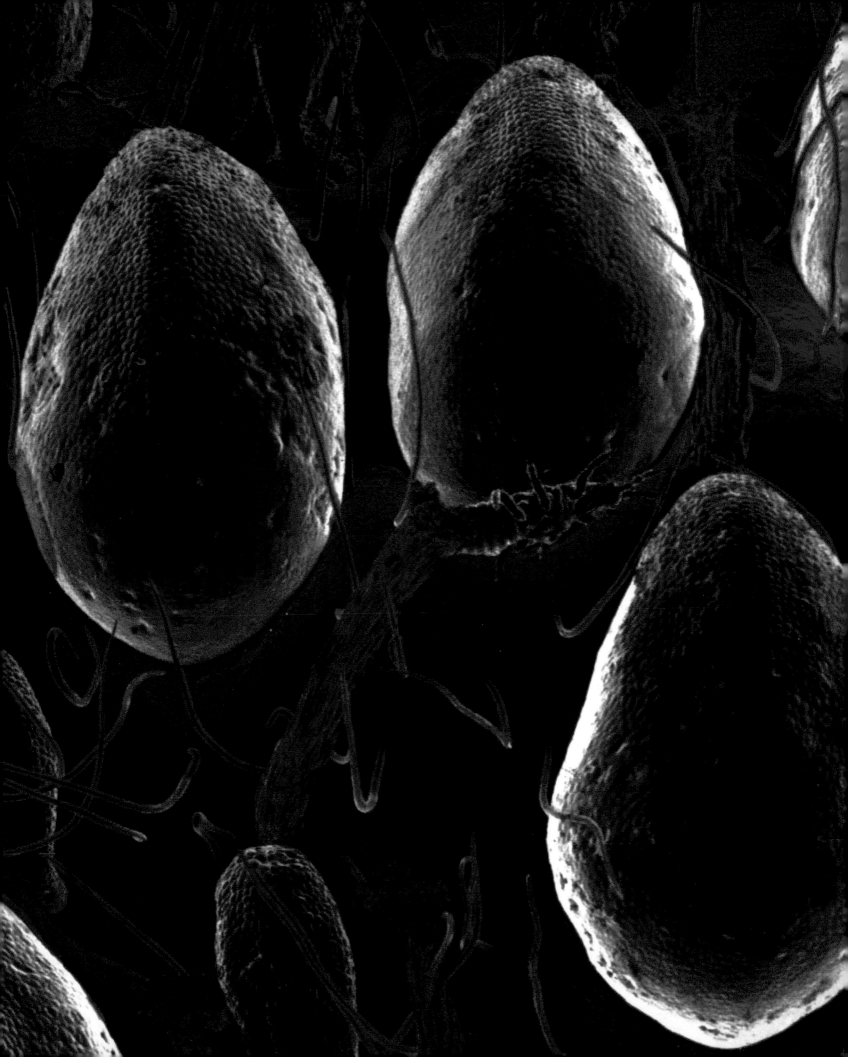

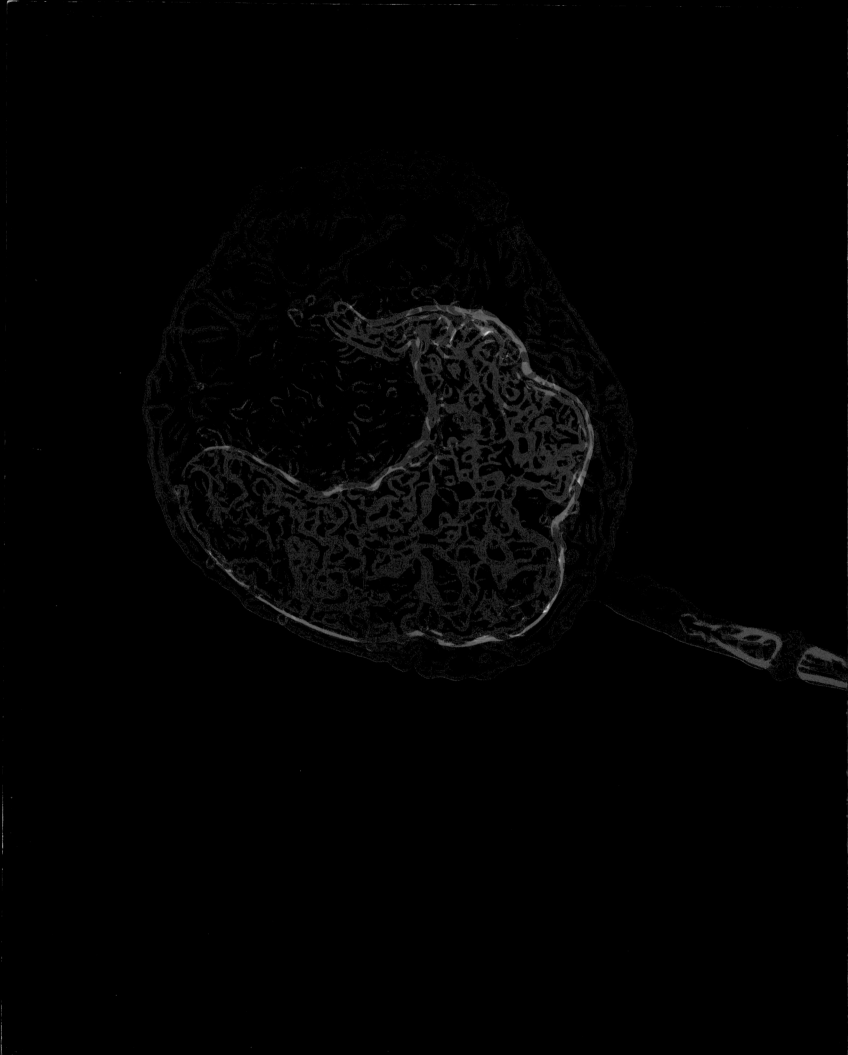

III

Human Body

HeLa cells

GLOWING COLORS HERE are due to fluorescent dyes placed into these cancer cells, then excited with a laser. The cell nuclei are blue, mitochondria are green and the structural protein actin is red. HeLa cells were established in 1952 as the first human cell line to be grown purely in the laboratory. They thrive unusually well in the laboratory and are used widely in virology (viral research) and the study of cell division.

CONFOCAL LIGHT MICROGRAPH OF HELA CANCER CELLS STAINED TO REVEAL THEIR STRUCTURE

2580X

HeLa cells

FLUORESCENT DYES TAGGED to structures within each cell reveal nuclei containing the genetic material chromatin (red). The yellow and white strands are the actin and microtubule cytoskeleton by which a cell can form rigid structures when required. HeLa cells, the first human cell line, were obtained from the cervix of Henrietta Lacks (after whom the cells are named) from Baltimore, USA, in 1951, who died of cervical cancer eight months later. They are still used in research worldwide.

CONFOCAL LIGHT MICROGRAPH OF HELA CANCER CELLS STAINED TO REVEAL THEIR STRUCTURE

1475

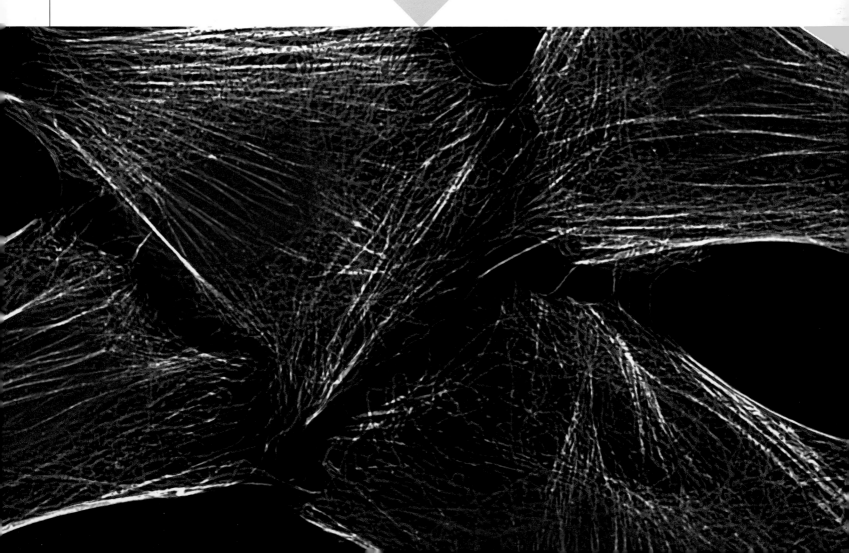

Mitosis

DURING MITOSIS TWO DAUGHTER NUCLEI (green) are formed from one parent
nucleus. At metaphase, chromosomes (yellow) line up along the center
of the cell, and the spindle fibers (light blue) grow from the poles (at
top and bottom) to the center of each chromosome. Chromosomes are
made up of two identical sister chromatids, which are separated into the
two daughter nuclei, so that each daughter cell retains the parent cell's
genetic information. The rounded nuclei of other cells are seen, as well
as the microfilaments that make up the cytoskeleton.

IMMUNOFLUORESCENCE LIGHT MICROGRAPH OF A CELL (CENTER LEFT)
IN METAPHASE DURING MITOSIS (NUCLEAR DIVISION)

1350✕

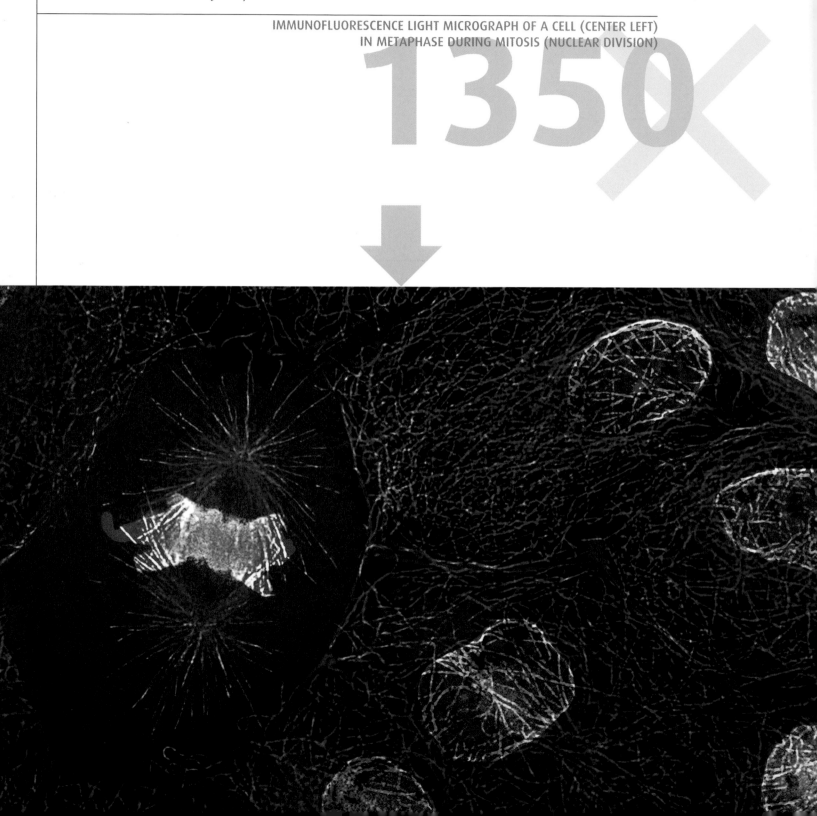

Cell division

CYTOKINESIS OCCURS after nuclear division (mitosis), which produces two daughter nuclei containing deoxyribonucleic acid (DNA, blue). A contractile ring made of actin filaments forms beneath the cells plasma membrane. This contracts, creating a furrow in the cytoplasm. The furrow deepens, separating the cytoplasm and cell organelles, until the two new cells are formed. The red structures are microtubules, part of the cell's cytoskeleton, along with the actin filaments. They separate the chromosomes during mitosis.

IMMUNOFLUORESCENCE LIGHT MICROGRAPH OF AN ANIMAL CELL DURING CYTOKINESIS, THE LAST STAGE OF CELL DIVISION

7300X

Mitochondria

FOUND IN ALL EUKARYOTIC CELLS, mitochondria are the site of cell respiration: the chemical process which uses molecular oxygen to oxidize sugars and fats to produce energy. The energy is stored as adenosine triphosphate (ATP) and is used by the cell to drive chemical reactions such as protein formation. Mitochondria have a double membrane (seen here) with the inner membrane folded to produce ingrowths called cristae. Other cell organelles are visible in the surrounding cytoplasm.

TRANSMISSION ELECTRON MICROGRAPH OF A MITOCHONDRION (CENTER)

115000✕

Stem cells

STEM CELLS FROM THE RETINA OF THE EYE are able to differentiate into any of the different retinal cell types. The type of cell they mature into depends upon the biochemical signals received by the immature cells. This ability makes retinal stem cells a potential source of cells to repair damaged retinal tissue and restore eyesight in diseases such as diabetic retinopathy and macular degeneration. The patient's own stem cells could be used and so would not be rejected.

TWO HUMAN RETINAL STEM CELLS (PURPLE AND PINK) ON THEIR FEEDER CELLS (GREEN)

6200

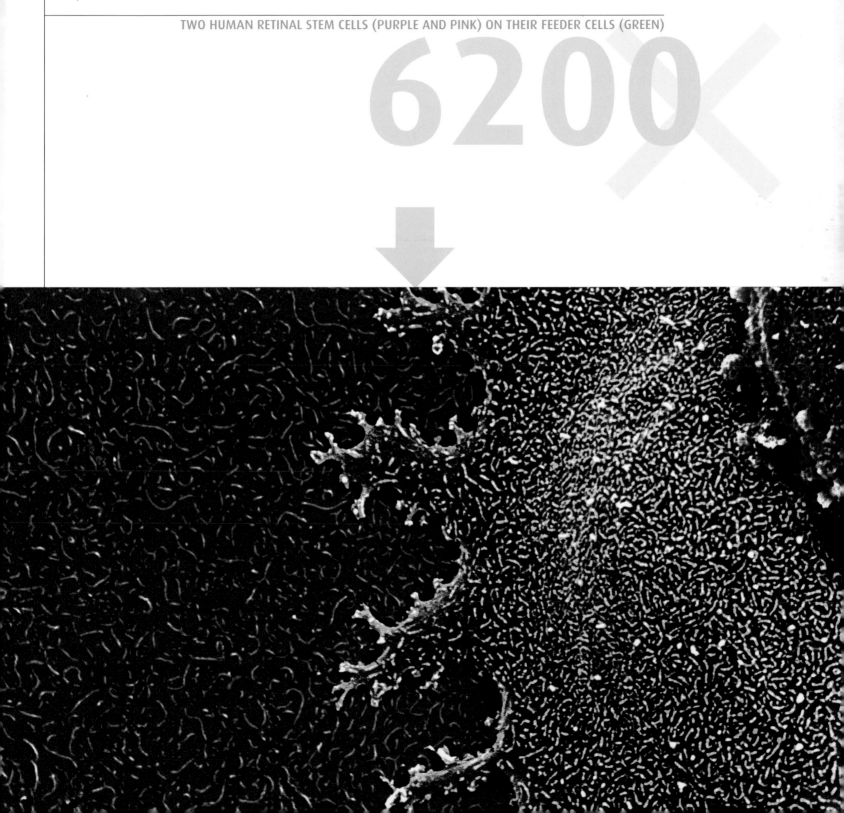

Nerve cell growth

THE CELL BODIES, typical of nerve cells, contain the nuclei (pink) and these initially spherical tumor cells have formed long branching extensions called neurites (yellow and blue). These neurites would form the axons and dendrites that normally connect nerve cells and transmit impulses around the body, the brain and the spinal cord. PC12 cells are research cells derived from a tumor of the adrenal gland (pheochromocytoma).

IMMUNOFLUORESCENCE LIGHT MICROGRAPH OF PC12 CELLS FOLLOWING
STIMULATION BY NERVE GROWTH FACTOR

2250✕

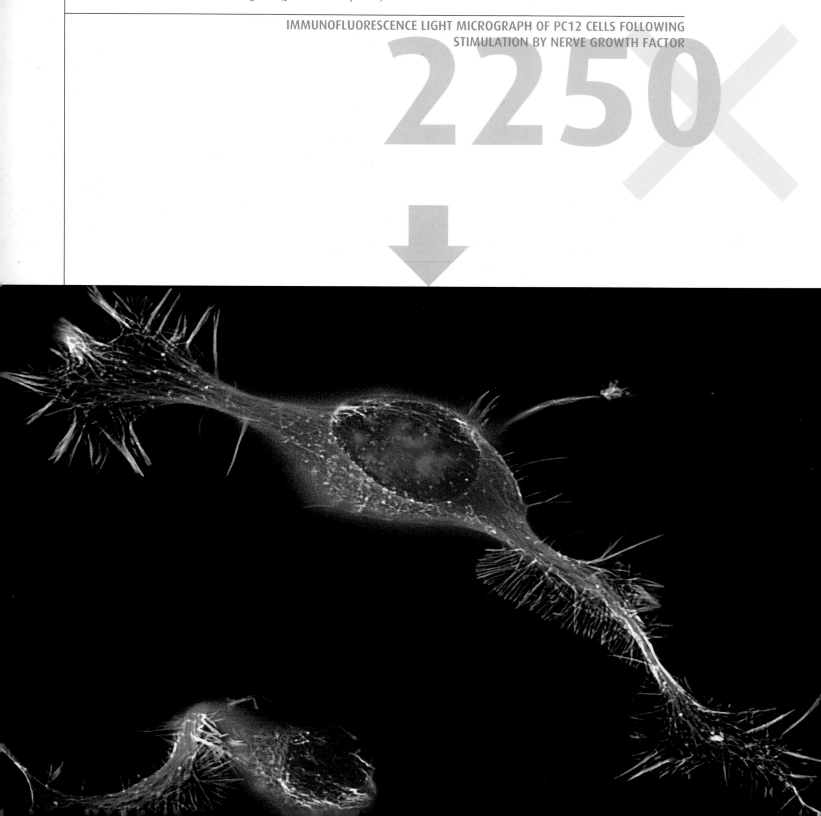

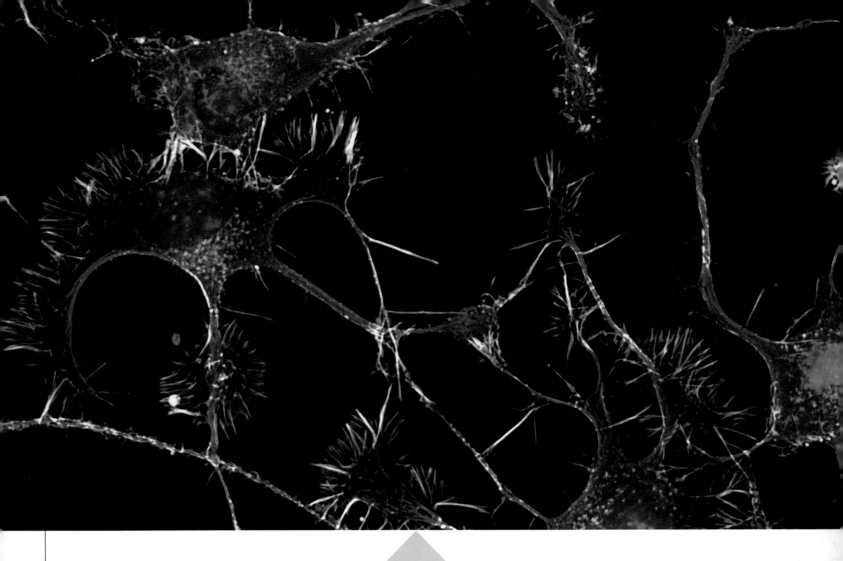

Nerve cell growth

THE LINKING UP of more PC12 culture cells is seen here, much like a natural network of nerve cells. Cell bodies contain the nuclei (pink), while the introduced nerve growth factor has produced these long branching extensions or neurites (yellow and blue). In proper nerve cells these neurites would be the axons and dendrites that transmit nerve impulses, but these PC12 cells are research cells. Research into curing spinal paralysis uses tissue cultures like this to investigate nerve regeneration.

IMMUNOFLUORESCENCE LIGHT MICROGRAPH OF PC12 CELLS FOLLOWING
STIMULATION BY NERVE GROWTH FACTOR

1540X

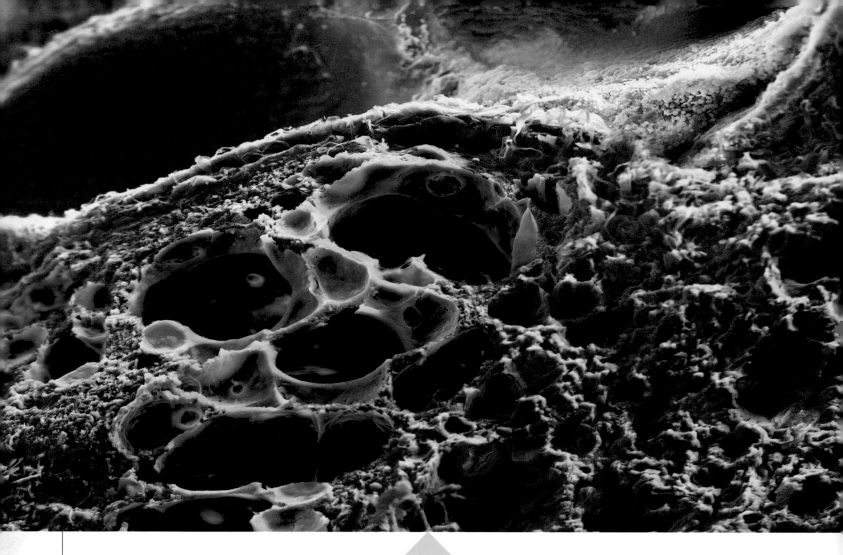

Lung bronchiole and alveoli

THE BRONCHIOLE is a small branching airway that carries air inhaled from the upper airways and windpipe to the tiny alveoli in the lungs. Each rounded alveolus is a site for gas exchange between the air in the air sac and the blood in adjacent capillaries in the alveoli walls. Oxygen diffuses into the blood for delivery to all cells in the body for respiration, while carbon dioxide diffuses out of the blood to be exhaled.

WHITE BRONCHIOLE (AT TOP) AND A CLUSTER OF ALVEOLI (AIR SACS, GRAY) IN A LUNG

500✕

Macrophage in the lung

ALVEOLI ARE THE TINY AIR SACS that are the sites of gaseous exchange in the lungs. Macrophages are a type of large white blood cell of the immune system that are found in tissues of the body rather than circulating in the blood. They recognize foreign particles, including bacteria, pollen and dust, and phagocytose (engulf) and digest them.

A MACROPHAGE IMMUNE CELL (ROUND) IN AN ALVEOLUS

2280 ✕

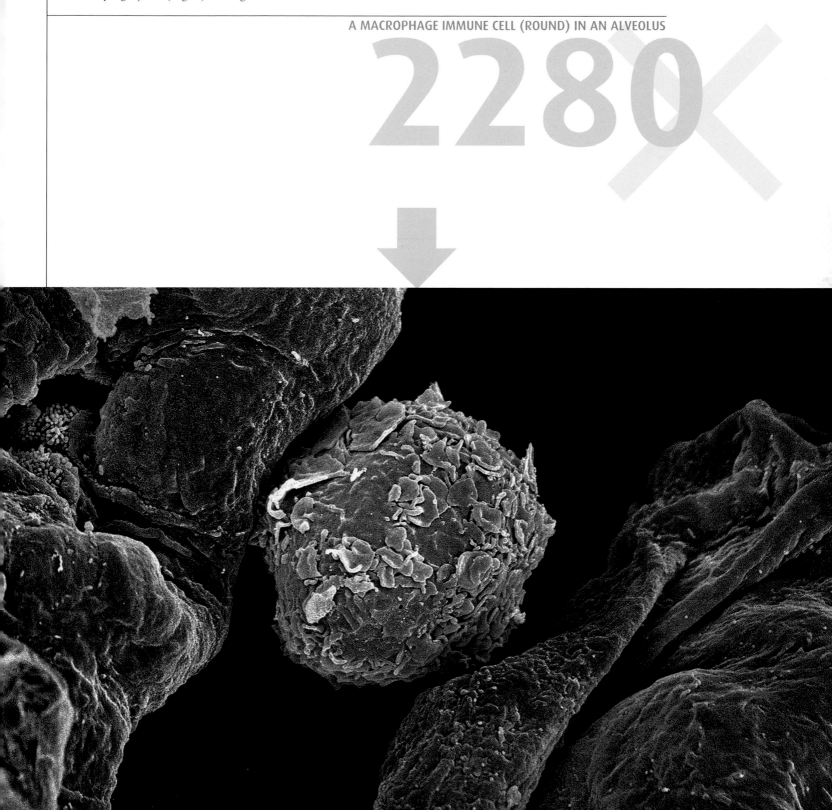

Fetal spinal column

BONE DEVELOPMENT OCCURS in the fetus from an early age and as early as twelve weeks old movements of the fetus begin. Here the developing spine running from upper left to right shows oval vertebrae (spongy, pink) separated by smaller intervertebral discs. The discs are made of fibrous cartilage, and are used for shock absorption, whilst the many vertebrae allow the spine the flexibility to move.

LONGITUDINAL SECTION THROUGH THE DEVELOPING SPINE OF A FETUS

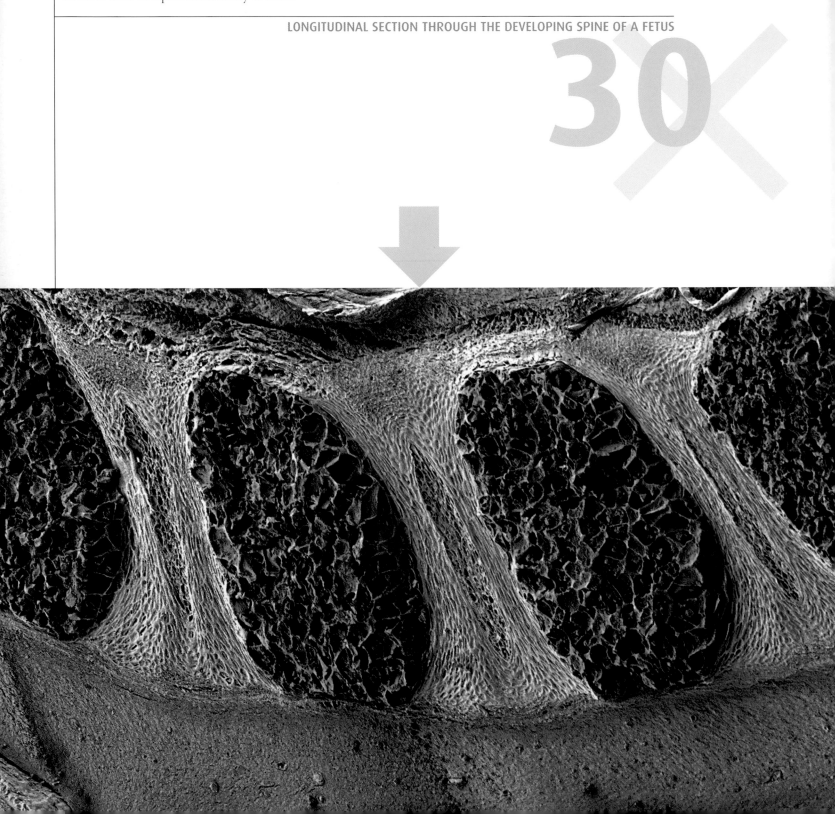

Bone tissue

BONE TISSUE CAN BE either cortical (compact) or cancellous. Cortical bone usually makes up the exterior of the bone, while cancellous bone is found in the interior. Cancellous bone has a spongy appearance comprising a network of trabeculae (rod-shaped) fibrous tissue. These structures provide support and strength to the bone. The spaces within this tissue contain bone marrow, a blood forming substance.

CANCELLOUS (SPONGY) BONE IN A HONEYCOMB ARRANGEMENT

24

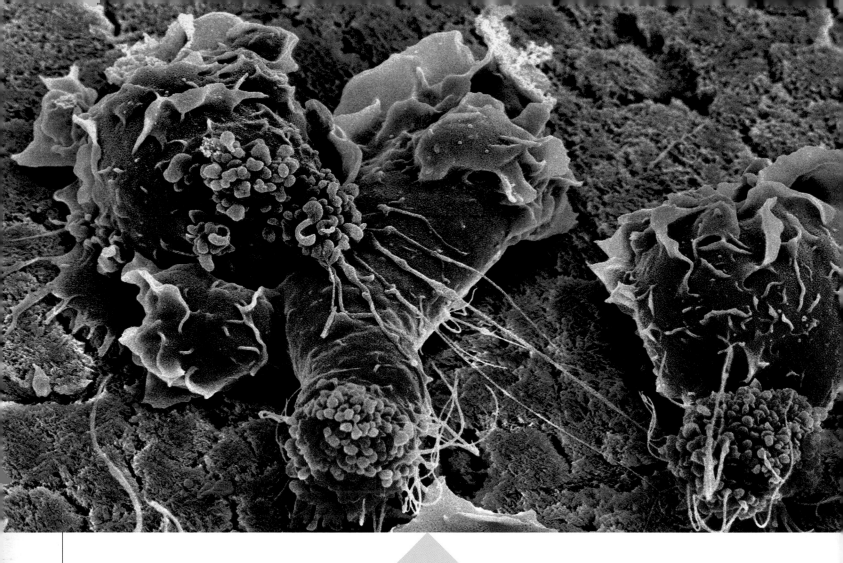

Bone cancer

THESE CELLS (BLUE) are monocytes, a type of white blood cell which enter body tissues where they differentiate into macrophage cells. In time, these will join together into a large, multinucleated cell called an osteoclast. Osteoclasts are normally present in bones to absorb and remove unwanted bone tissue during normal bone regeneration but they can become cancerous. They develop into an osteoclastoma or giant cell tumor, a type of bone cancer that usually affects the ends of the long bones.

BONE CANCER PRECURSOR CELLS ON THE SURFACE OF A BONE

8200X

Cardiac muscle

THE MUSCLE FIBERS RUN from left to right, and are made up of numerous myofibrils (not clearly seen). The fibers are crossed by transverse tubules (lines running vertically). These tubules mark the division of the myofibrils (within the fibers) into contractile units known as sarcomeres. Cardiac muscle, mostly under subconscious control, continuously contracts and pumps blood around the body without tiring.

FREEZE-FRACTURED SECTION THROUGH HEALTHY HEART (CARDIAC) MUSCLE FIBERS

12800✗

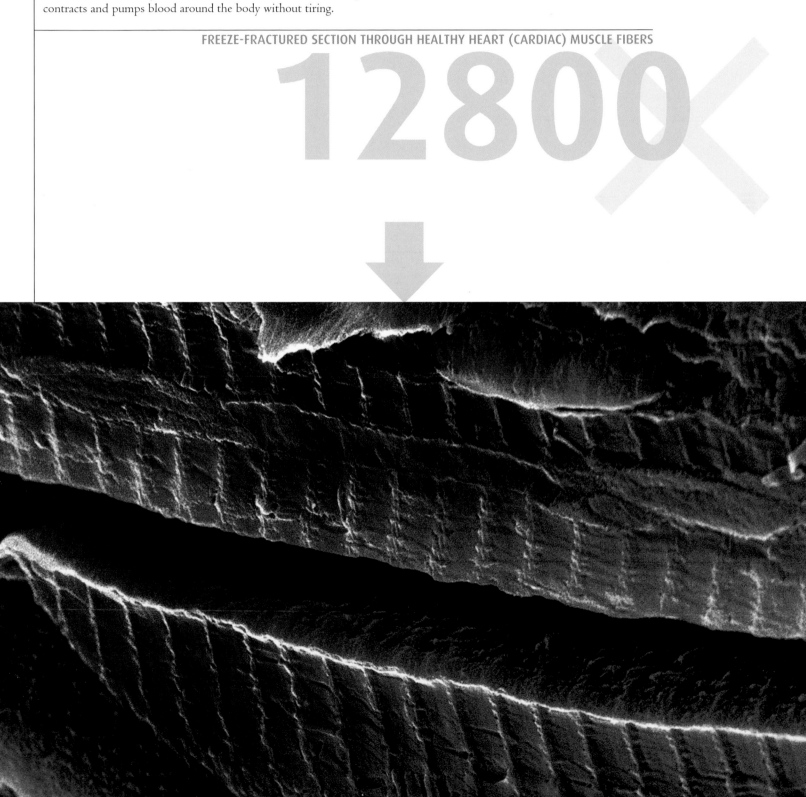

Smooth muscle fibers

THE FRACTURE PLANE CUTS across the smooth muscle fibers to reveal the smaller
fibers within the larger fibers. Smooth muscle is found in places in the body
like the digestive system and in blood vessels. It is not under voluntary control,
instead being controlled by the autonomic nervous system are by hormones.
Smooth muscle produces slow, long-term contractions.

FREEZE-FRACTURED BUNDLE OF SMOOTH MUSCLE FIBERS (BROWN),
WITH A BLOOD VESSEL (PINK, AT TOP)

8450X

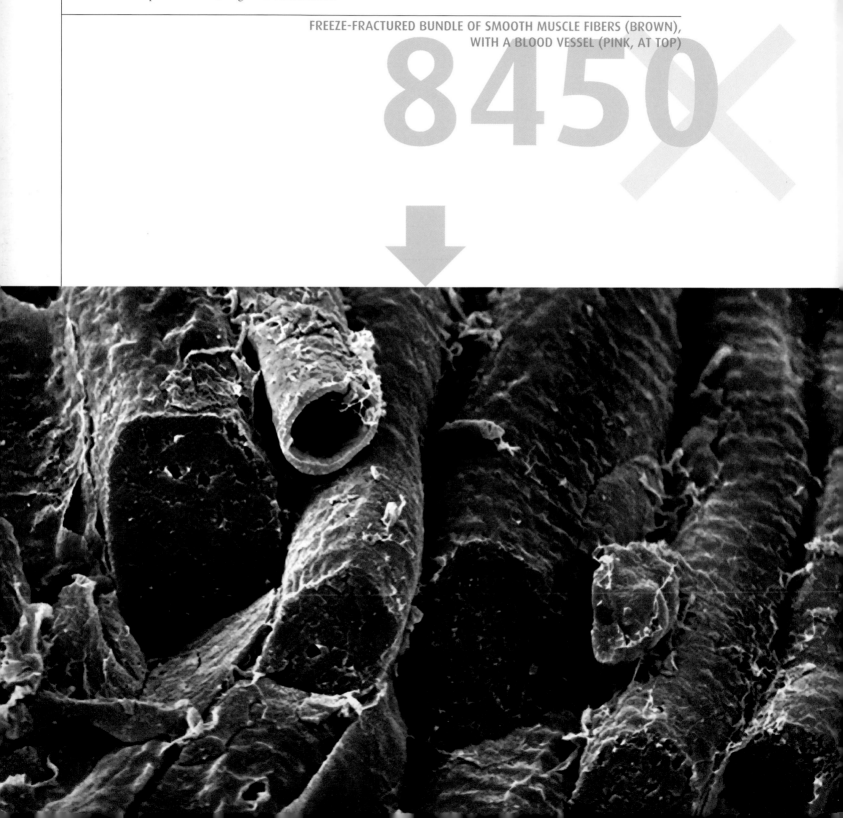

Capillaries on muscle

THE CAPILLARY NETWORK SUPPLIES the muscle with oxygen and nutrients, which it uses to create energy by respiration. The blood also takes away the waste product of respiration, carbon dioxide. The capillaries are so fine that a red blood cell (erythrocyte) can only just pass through.

CAPILLARIES (RED) CARRYING BLOOD TO MUSCLE FIBERS (BROWN)

1500✗

Blood vessels

THIS NETWORK OF BLOOD VESSELS infiltrates the tissue, supplying it with blood. Gases and nutrients are exchanged between the blood and surrounding tissue through the permeable walls of capillaries, the smallest blood vessels. The cast was made by injecting resin into the blood vessels. The surrounding tissues were then chemically dissolved.

RESIN CAST OF BLOOD VESSELS IN A LYMPH NODE

105✕

Brain blood vessel secretory cells

THE CHOROID PLEXUS is a network of capillary blood vessels found in each of the four ventricles (fluid-filled cavities) of the brain. The tips are swollen due to increased secretion of the cerebrospinal fluid which they produce. This cerebrospinal fluid surrounds and protects the brain and spinal cord.

FREEZE-FRACTURED CHOROID PLEXUS FROM A BRAIN. THE FRACTURE PLANE (GRAY) HAS REVEALED THE VERTICAL COLUMNS OF SECRETORY CELLS, WHOSE TIPS MAKE UP THE SURFACE (UPPER RIGHT)

16000X

Kidney stone crystals

KIDNEY STONES FORM when salt compounds precipitate out of the urine, typically as calcium oxalate. The resulting hard, round stones may cause severe pain and fever, particularly if they are large, as they pass down the urinary tract. They can also damage the kidney and impair its function. Large stones may need to be surgically removed or broken down using ultrasound waves.

THE CRYSTAL SURFACE OF A KIDNEY STONE (CALCULUS)

450×

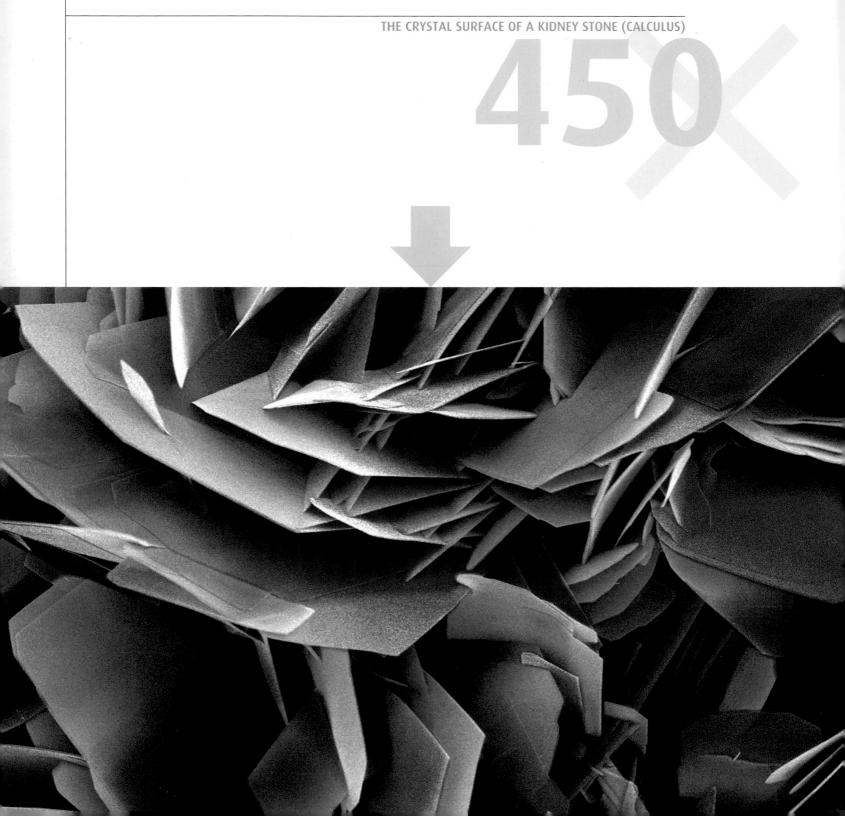

Blood clot crystals

WHEN THE SKIN IS CUT, small blood vessels are ruptured, releasing blood. Some proteins in the blood plasma (such as albumin) harden in the air to form crystals (pink) over the wound. Other blood proteins then help form a clot over the wound, preventing excessive blood loss, and keeping the wound free of bacteria and other foreign bodies.

CRYSTALS OF ALBUMIN, KNOWN AS HUMAN SERUM ALBUMIN, FROM A BLOOD CLOT

1060✕

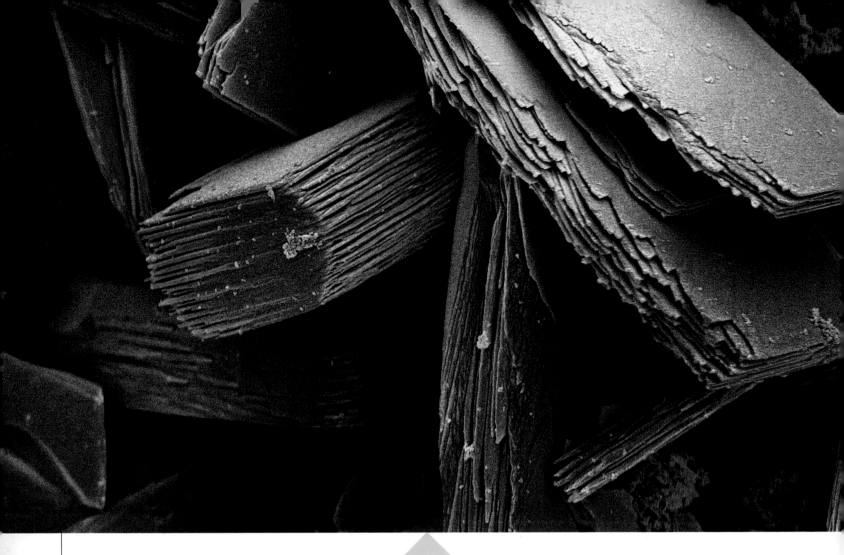

Blood clot crystals

THIS IMAGE AND THE PREVIOUS ONE neatly show the crystallizing effect as albumin in the blood comes into contact with air at an injury site on the body. Human serum albumin, as it is called, makes up 60% of the protein in human blood plasma and thus, as the most abundant plasma protein, helps to maintain blood volume. Albumin is produced in the liver and its other roles in the blood include helping to maintain the levels of hormones and calcium, and assisting water flow between the bloodstream and body tissues.

STACKED ALBUMIN CRYSTALS FROM A BLOOD CLOT SEALING OFF A WOUND

2400✕

Blood clot

RED BLOOD CELLS AND PLATELETS (smaller cells) trapped by a web of thin white strands of fibrin, an insoluble protein. Blood clotting is the solidifying of blood that occurs when blood vessels are damaged. This activates platelets in the blood which in turn help to stimulate the formation of fibrin filaments that enmesh the cells to form a solid clot. A clot may occur on the surface of the skin in the case of an injury, sealing the wound and preventing excessive loss of blood. A clot may also form dangerously inside a diseased blood vessel where it is called a thrombus and may cause a heart attack.

TANGLE OF THREADS AND RED BLOOD CELLS FORMING A BLOOD CLOT

5900✕

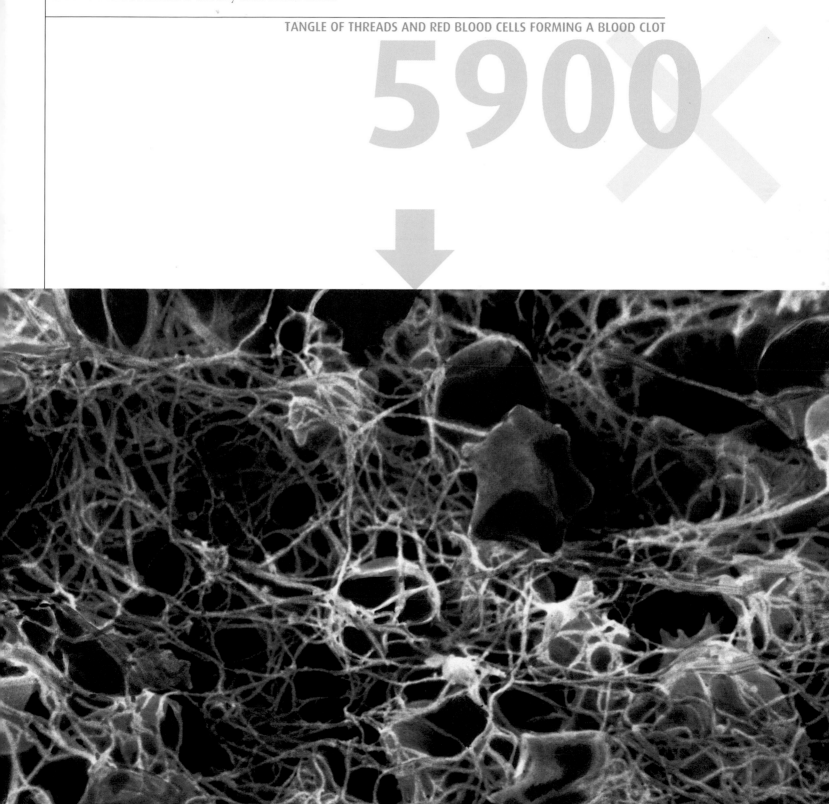

Sperm

PRODUCED BY THE TESTES these tiny male sex cells are responsible for fertilizing the female egg (ovum). Each sperm has a rounded head which contains the male hereditary material (DNA) and a long tail which helps it to swim through the female reproductive tract in search of the egg. While the female usually only produces one egg at a time, about 300 million sperm are ejaculated by the male, but only one sperm can fertilize the ovum.

CLUSTER OF HUMAN SPERM (SPERMATOZOA)

4480

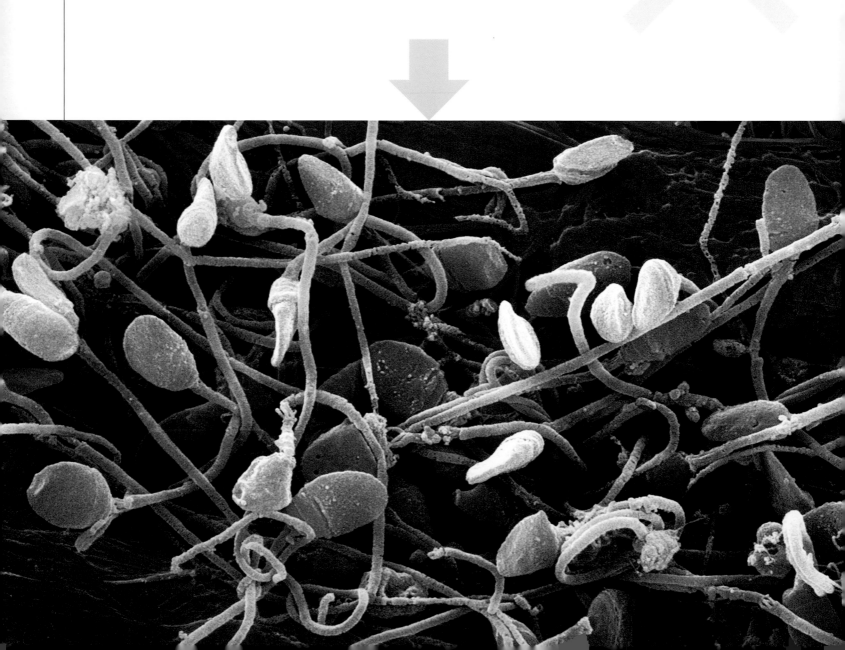

Fallopian tube

THE FALLOPIAN TUBES (OVIDUCTS) are the two channels through which
eggs pass from the ovaries into the womb. Fimbriae are the folds at the
opening of each fallopian tube, closely situated next to the ovary. With
tiny cilia beating on their folds they help to guide a newly released egg
(ovum) into the tube itself. If an egg encounters sperm in the fallopian
tube it may be fertilized, in which case it can implant into the wall of the
womb and pregnancy can begin.

FIMBRIAE OF A FALLOPIAN TUBE IN A WOMAN

47X

Chromosomes

EACH CHROMOSOME CONSISTS of two identical chromatids joined at a centromere. Chromosomes are only seen in this compact structure during cell division. They are found in every cell nucleus in the body and consist of genetic material (DNA) in the form of genes in association with proteins. Apart from the sex cells, the sperm and egg, all other human cells contain 46 chromosomes, with 23 inherited from the mother, and 23 from the father.

HUMAN CHROMOSOMES, THE INHERITED GENETIC MATERIAL

21000X

Eyelash hairs and skin

THE SHAFTS OF HAIR, made up of a fibrous protein called keratin, are anchored in their individual hair follicles in the surface of the skin. On the skin surface are keratinized dead cells that detach from the body naturally. These squamous (flattened) cells arise from the lower, living layers of skin. The tails of tiny eyelash mites (Demodex folliculorum) are seen protruding from the base of several eyelashes.

EYELASH HAIRS GROWING FROM THE SURFACE OF HUMAN SKIN

192

Nasal lining

THE NASAL CAVITY WARMS, humidifies and filters inspired air. The lining contains moisture- secreting cells, mucus-secreting cells and sensory cells. Mucus traps foreign particles such as bacteria and dust, preventing them from entering the lungs.

STRATIFIED SQUAMOUS EPITHELIUM CELLS FROM THE LINING OF THE NOSE

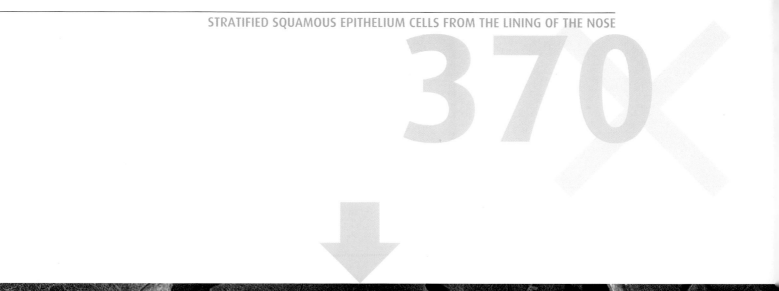

370

Iris cells of the eye

PIGMENT CELLS (melanocytes, blue and brown) can be seen here, joined loosely together by connective tissue fibers (white). Smaller macrophage immune cells dot the surface. Beneath this matrix (stroma) of iris cells lie muscle fibers. These muscle fibers contract and dilate reflexively, and so control light entering the central pupil of the iris, and falling on the retina. Mobility of the iris is possible through this loose network structure of cells.

SURFACE CELLS ON THE IRIS OF THE EYE, THE COLORED PART OF THE EYE

1550×

Eye lens surface

THE LENS, WHICH IS FOUND at the front of the eye, focuses light onto the light-sensitive cells in the retina at the back of the eye. Cuboidal epithelium is found at the front of the lens.

FREEZE-FRACTURED LENS (CRYSTALLINE) OF AN EYE, SHOWING THE CUBOIDAL EPITHELIUM CELLS (LAYERS)

345X

Lens cells of the eye

THE TRANSPARENCY OF THE LENS is due to an absence of nuclei in its cells and to the crytalline precision of their arrangement. The zip-like rows of ball-and-socket joints that bind these cells together may play a part. In cross-section lens cells are flattened hexagons, arranged in regular stacks. These cells are called fibers because of their greatly elongated form.

FREEZE-FRACTURED LENS OF THE HUMAN EYE, SHOWING THE ORDERED
ARRANGEMENT OF THE CLOSELY PACKED LENS CELLS CALLED FIBERS

1650×

Eye blood vessels

THE CHOROID IS A TISSUE LAYER beneath the retina which supplies blood, thus food and oxygen, to light sensitive cells of the eye. Many cone cells occur on the retina at the central fovea area, specialized for acute day vision, and hence an area that uses much blood. The choroid is also darkened (blue) with pigment cells; pigment absorbs light rays passing through the retina and prevents light reflection.

BLOOD VESSELS IN THE CHOROID OF THE EYE. A BRANCHING NETWORK OF ARTERIES AND VEINS CAN BE SEEN IN THIS AREA UNDER THE CENTRAL FOVEA

150×

Ciliary processes of the eye

CILIARY PROCESSES (BLUE) form a series of about 70 radial ridges (eight seen here) arranged around the eye lens. (not seen, off bottom). Zonular fibers (yellow) arise from the ciliary processes and attach to the equatorial region of the eye lens. Ciliary muscles (not seen) act on these ciliary processes to control the shape of the lens and so focus images onto the retina. Remnants of the basal vitreous body (pink) are at top.

INNER VIEW OF THE CILIARY PROCESSES OF THE EYE, THE STRUCTURES WHICH HOLD THE EYE LENS

182X

Ciliary body

THE CILIARY BODY (across upper frame) forms a ring between the iris (at bottom and off frame) and the choroid, the inner surface of the eyeball (at top). It is the colored iris that surrounds the pupil through which light shines into the eye. The ciliary body joins to ligaments that hold the lens in place behind the iris. The lens has been removed here. The ciliary body also contains the ciliary muscle that is contracted to alter the curvature of the lens and focus light on the retina.

SECTION THROUGH AN EYE TO SHOW A PARTIAL VIEW OF THE STRUCTURES OF THE FRONT OF THE EYE, AS SEEN FROM WITHIN THE EYEBALL

Choroid layer of the eye

A SINGLE PIGMENTED epithelial cell with nucleus (red) and granular contents is seen within the choroid. The choroid layer is the inner surface of the eyeball, lying behind the white of the eye (sclera) and behind the retina, the part of the eye containing light-sensitive cells. It is pigmented to absorb excess light, preventing internal reflections that would form multiple images on the retina.

SECTION THROUGH THE CHOROID LAYER OF THE EYE SHOWING A PIGMENT CELL

2475✕

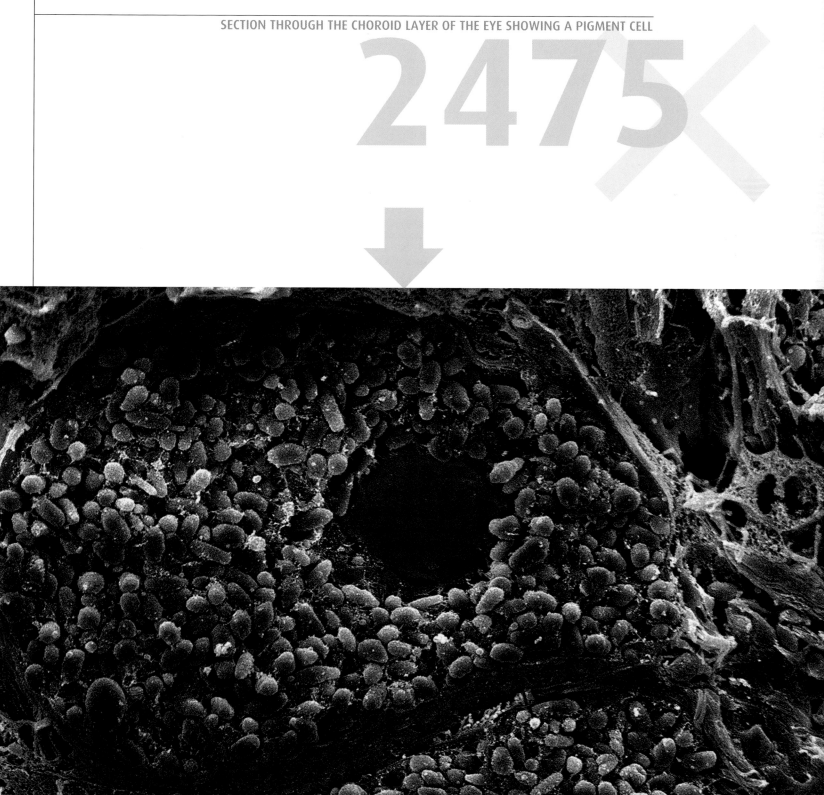

Sensory hair cells in the ear

THE ROWS OF CRESCENT-SHAPED STRUCTURES (across center) are numerous stereocilia, and are located on top of supporting hair cells. Sound waves entering the inner ear displace the fluid that surrounds the stereocilia, causing them to bend. This triggers a response in the hair cells, which release neurotransmitter chemicals that generate nerve impulses. The nerve impulses travel to the brain along the auditory nerve. This process can transmit information about the loudness and pitch of a sound.

HAIR CELLS IN THE COCHLEA, THE INNER EAR'S AUDITORY SENSE ORGAN

4600✕

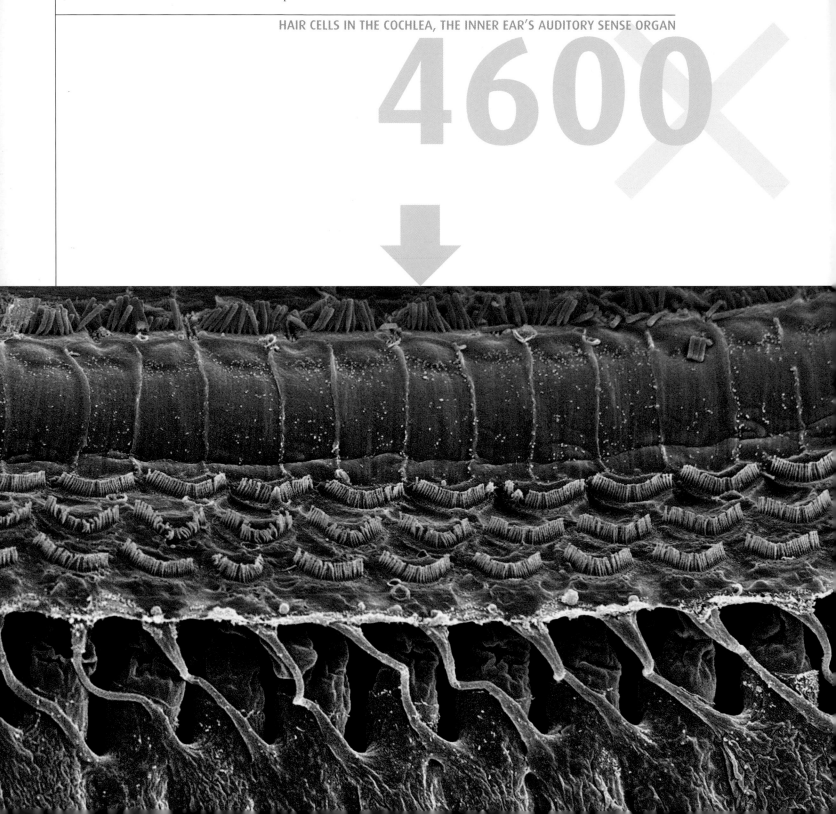

Tongue

SLENDER AND SPIKY FILIFORM PAPILLAE (red) are the most common type, while fungiform papillae (purple) are rounded and flat. Filiform papillae form a rough surface which aids in feeding. Taste buds are located around the base of many of the larger fungiform papillae. The untidy appearance of this surface is due to dead cells which are constantly being shed from the top layer of the tongue.

DORSAL SURFACE OF THE TONGUE COVERED IN TWO TYPES OF PAPILLAE (PROJECTIONS)

200X

Tongue papillae

DUE TO THE ROLE the tongue plays, dead cells from the uppermost epithelial layer of cells are naturally shed and regularly replaced (desquamation) and this gives the papillae a scaly appearance. Filiform papillae are the most numerous on the tongue, contain nerve endings for touch sensations, and help the tongue to manipulate food around the mouth.

FILIFORM PAPILLAE (CONE-SHAPED) ON THE SURFACE OF THE TONGUE

650x

Esophagus lining

RESEMBLING A GARDEN MAZE, this network of tiny ridges (microplicae) decorating the surface of epithelial cells helps to protect the lining of the esophagus from food abrasion. These microplicae also prevent the epithelium from drying out; they trap surface secretions (mainly mucus) within their folds, to keep the epithelium moist and lubricated. The esophagus is a muscular tube, about 10 inches (25 cm) long, that runs from the back of the throat down to the stomach.

LINING OF THE ESOPHAGUS (GULLET) WITH BACTERIA ON THE SURFACE

9500✗

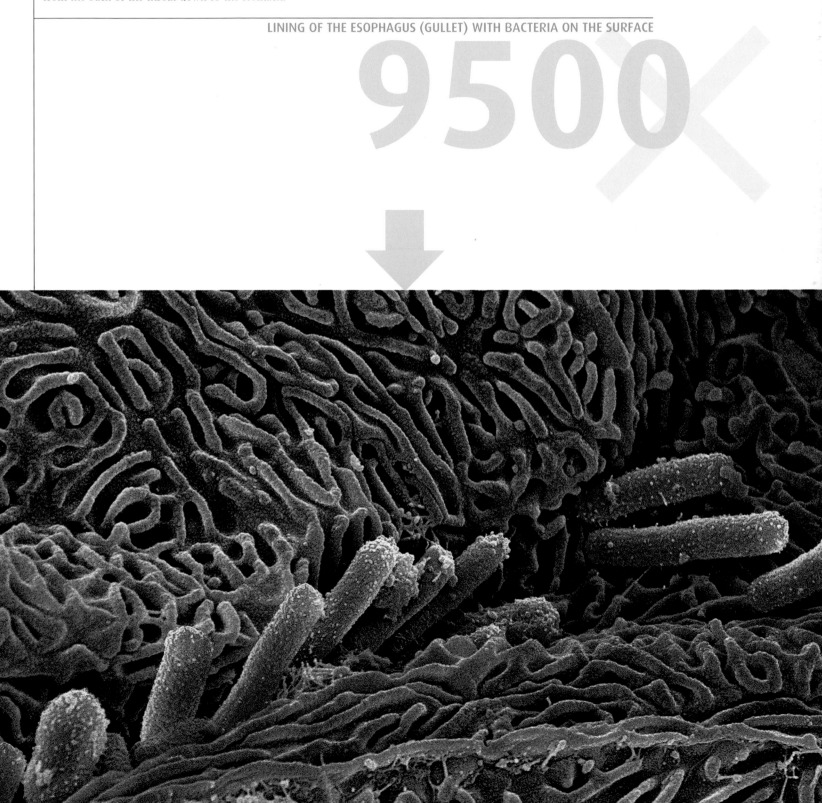

Esophagus lining

THE LINING OF THE ESOPHAGUS is composed of squamous epithelial cells. On the surface of these cells are tiny ridges (microplicae) which protect against the abrasion of rough food and help the food to pass. Many types of bacteria are found in the alimentary canal, of which the esophagus is the upper part, and most of these bacteria serve an important function in keeping the gut healthy. Rod-shaped varieties of bacteria are known as bacilli.

LINING OF THE ESOPHAGUS (GULLET) WITH BACTERIA ON THE SURFACE

2500✕

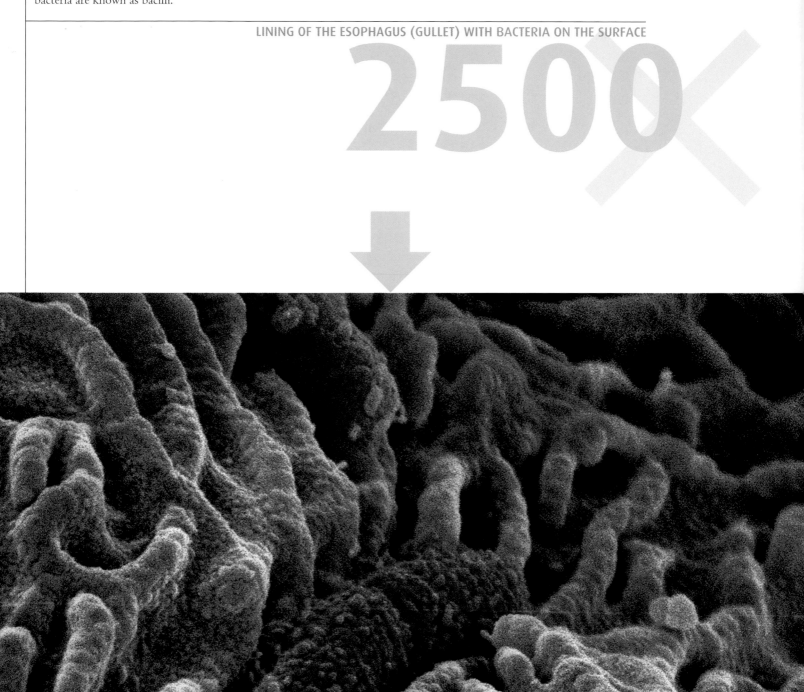

Duodenal villi

VILLI ARE FOLDS IN THE DUODENAL WALL, which project into the lumen. These folds greatly increase the absorptive and secretory surface of the duodenum. Digestive and stomach-acid neutralizing enzymes are secreted by the duodenum, and food is further digested here. Nutrients from the food are then absorbed into the blood through capillaries which are contained in each villus. To aid food absorption, villi are more widely distributed in the duodenum than elsewhere in the intestine.

VILLI IN THE DUODENUM, THE UPPER PART OF THE SMALL INTESTINE

480✕

Bile duct

THE BILE DUCTS TRANSPORT BILE, a dark fluid produced by the liver, to the gall bladder where it is stored before being passed on to the small intestine where it helps in the digestion of fats. The open lumen (interior) of this tube through which bile flows is seen here with a flattened shape. Lined by a thin layer of columnar epithelium (yellow), the bile duct also consists of connective tissue which provides structural support and blood to the vital epithelial layer.

CROSS-SECTION THROUGH A BILE DUCT AT LOW MAGNIFICATION

2K

Bile duct

THE INNER LINING of the bile duct consists of a row of columnar epithelial cells (yellow, running from lower left to right) with the undulating duct surface seen at top. These epithelial cells have microvilli, tiny finger-like projections that increase their surface area in the lumen. This allows the cells to reabsorb water and small molecules from the bile, thus concentrating the fluid. Below the epithelial layer is the lamina propria (brown, bottom), a type of connective tissue that is highly vascular and supplies blood to the ducts. Bile is an alkaline liquid that aids the digestion and absorption of fat and fat-soluble vitamins in the small intestine.

FREEZE-FRACTURED SURFACE LAYERS OF A BILE DUCT

245X

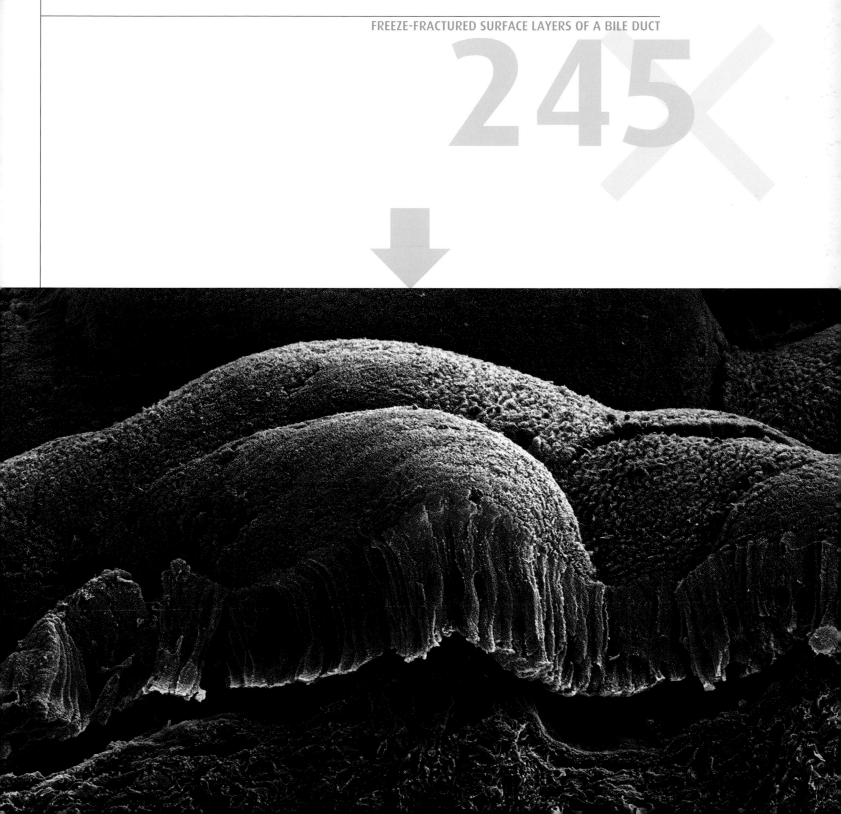

Intestinal surface

VILLI GREATLY INCREASE the intestinal surface area for absorbing nutrients from food. The epithelium (outer layer, folded) of a villus contains enterocyte cells, which are involved in nutrient absorption, and goblet cells, which secrete mucus onto the intestinal surface. The inner villus tissue, that is exposed by this section, will contain blood vessels for the transport of digestive products to a nearby vein.

SECTIONED VILLUS (FOLD) ON THE LINING OF THE SMALL INTESTINE

670X

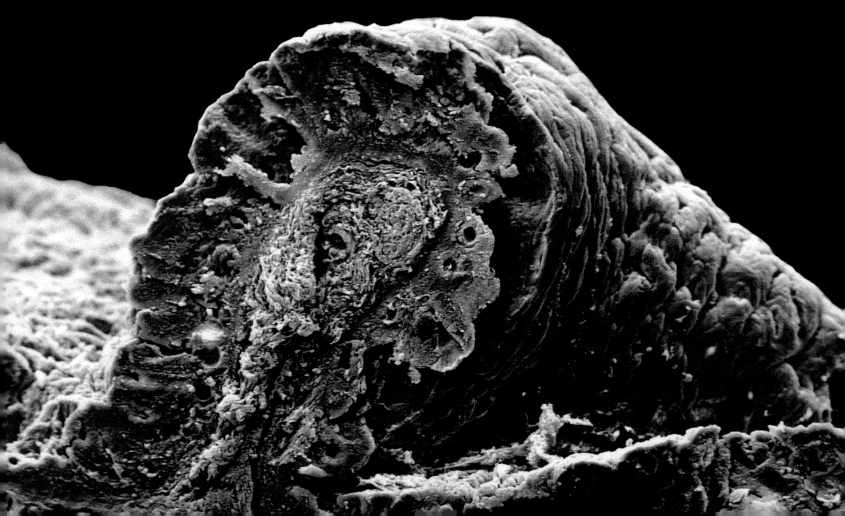

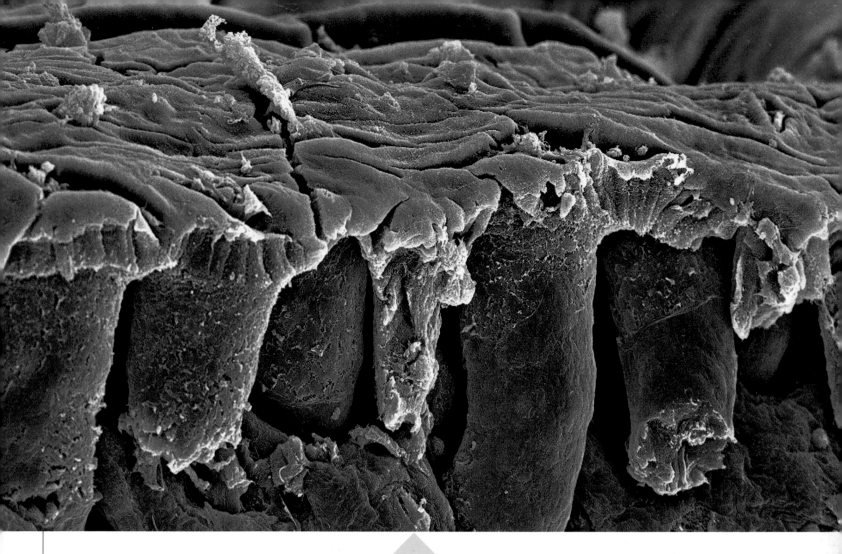

Large intestine lining

THE INTERNAL SURFACE (lumen, at top) of this region of the gut is where water, minerals and other molecules are absorbed from the remains of food debris. This surface is lined with a columnar epithelium, a thin layer of cells running across the center from left to right. Beneath the epithelium are larger tubular intestinal glands (crypts of Lieberkuhn), appearing here as five big cylindrical structures. Their primary purpose is in secreting a mucus which lubricates the feces passing over the surface.

FREEZE-FRACTURED SECTION THROUGH PART OF THE SURFACE OF THE LARGE INTESTINE (COLON)

530X

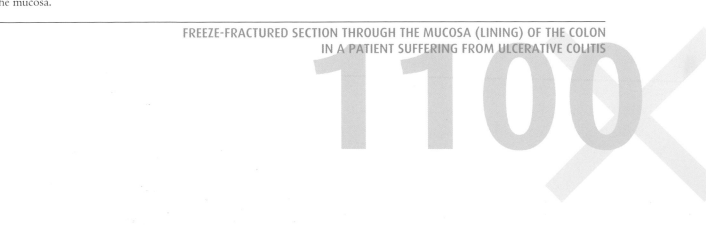

Ulcerative colitis

THIS CONDITION CAUSES SEVERE PAIN and inflammation of the lining of the colon (large intestine) and rectum. The mucosa (dark pink, at center) contains goblet cells which secrete lubricating mucus (pink, upper frame) onto the surface of the intestine (white). The numbers of goblet cells present are reduced in this disease making the surface more susceptible to ulcers. Red blood cells (red) can be seen in the vascular tissue beneath the mucosa.

FREEZE-FRACTURED SECTION THROUGH THE MUCOSA (LINING) OF THE COLON
IN A PATIENT SUFFERING FROM ULCERATIVE COLITIS

1100×

Purkinje nerve cells

THE CEREBELLUM OF THE BRAIN is composed of outer layers of gray matter and an inner core of white matter made up of nerve fibers. Within the gray matter are Purkinje nerve cells (two large bodies) from which branch many thread-like dendrites and a thicker axon. The dendrites relay nerve impulses to the Purkinje cell body and in this way each cell shares many interconnections and information with other nerve cells. Purkinje cells are among the largest neurones (nerve cells) in the body.

TWO PURKINJE NERVE CELLS FROM THE CEREBELLUM OF THE BRAIN, THE PART OF THE BRAIN THAT CONTROLS BALANCE, POSTURE AND MUSCLE CO-ORDINATION

2875✕

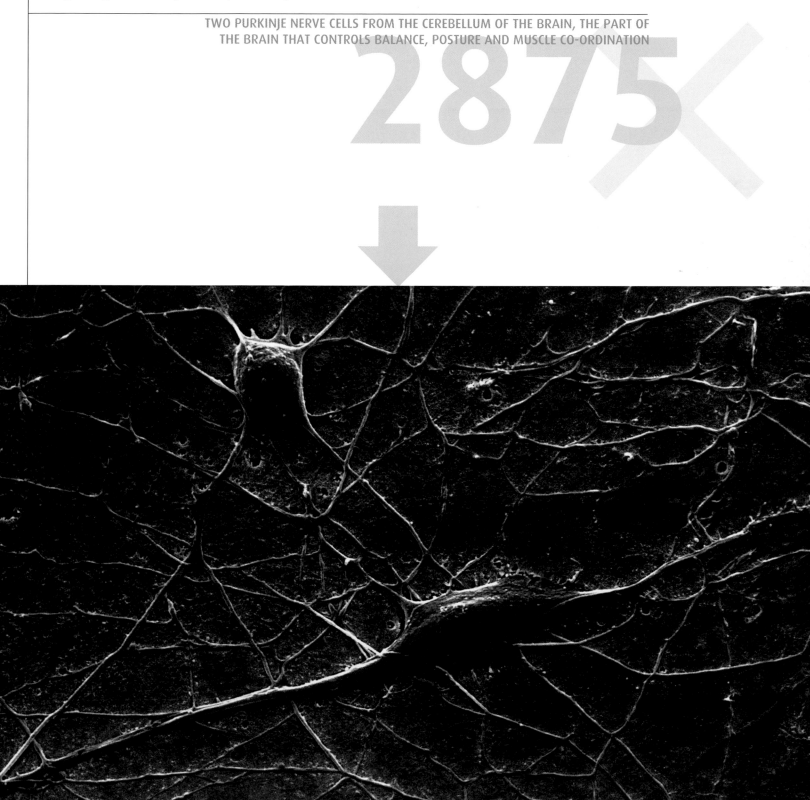

Myelinated nerve fiber

AT CENTER, A NERVE CELL AXON (red) is the structure through which nerve impulses are transmitted from a nerve cell. In a myelinated nerve, the fiber is surrounded by an insulating fatty layer (myelin sheath, brown). Myelin isolates the nerve fiber from its surroundings and helps to increase the speed of nerve transmission. Nerves serve to collect, interpret and relay information, passing it to other parts of the body, such as the muscles and organs like the brain.

CROSS-SECTION THROUGH A NERVE FIBER TO REVEAL A MYELIN SHEATH

28000×

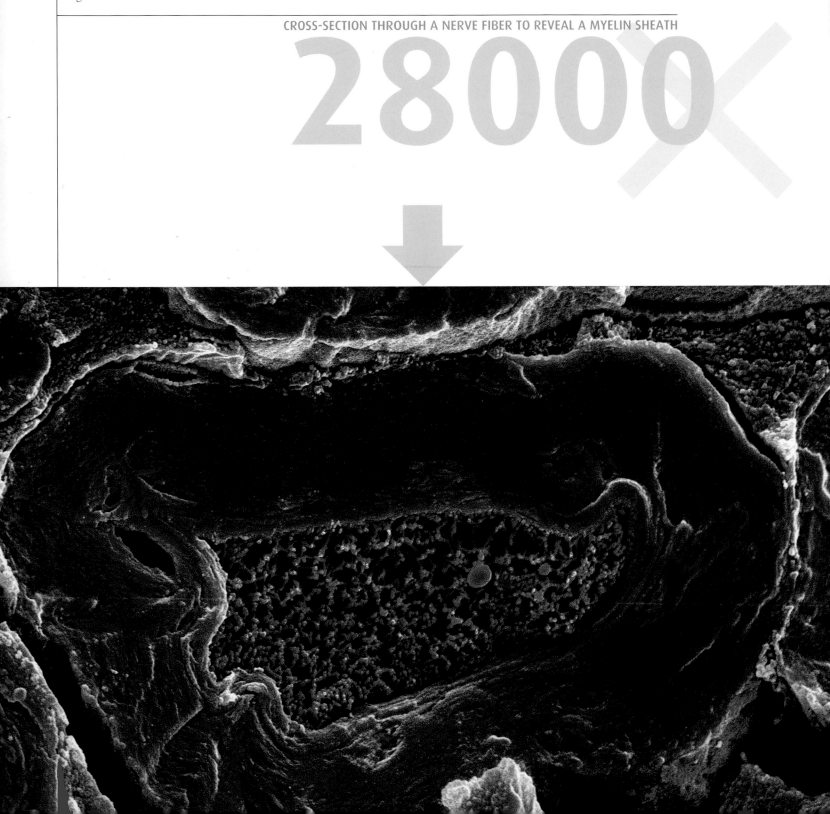

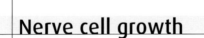

Nerve cell growth

THE FLUORESCENT STAINS SHOW NEURITES (thin strands, either axons or dendrites) connecting nerve cell bodies. Short neurites are green, long ones are red. Two large clusters of nerve cell bodies are at left and right, with non-neuronal cell nuclei in the background. Neurons transmit electrical signals around the body, especially the brain and the spinal cord. These neurons have been grown from NT2 cells, a human teratocarcinoma cell line that can differentiate into nerve cells. Such research may make neural regeneration treatments possible.

IMMUNOFLUORESCENCE LIGHT MICROGRAPH OF NERVE CELLS (NEURONS) GROWN IN CULTURE

320X

Nerve cell on a silicon chip

THE NERVE CELL WAS CULTURED on the circuit until it formed a network with nearby neurons (for example, far left). Under each cell is a transistor, which can excite the neuron above it. The neuron then passes a signal to the neurons attached to it, which activates the transistors beneath them. This experiment shows that hybrid neuron-silicon circuits are feasible. This research was conducted at the Max Planck Institute, Germany.

NEURON (NERVE CELL) ON A SILICON CHIP. THE NEURON IS IMMOBILIZED BETWEEN POLYIMIDE PILLARS

1650X

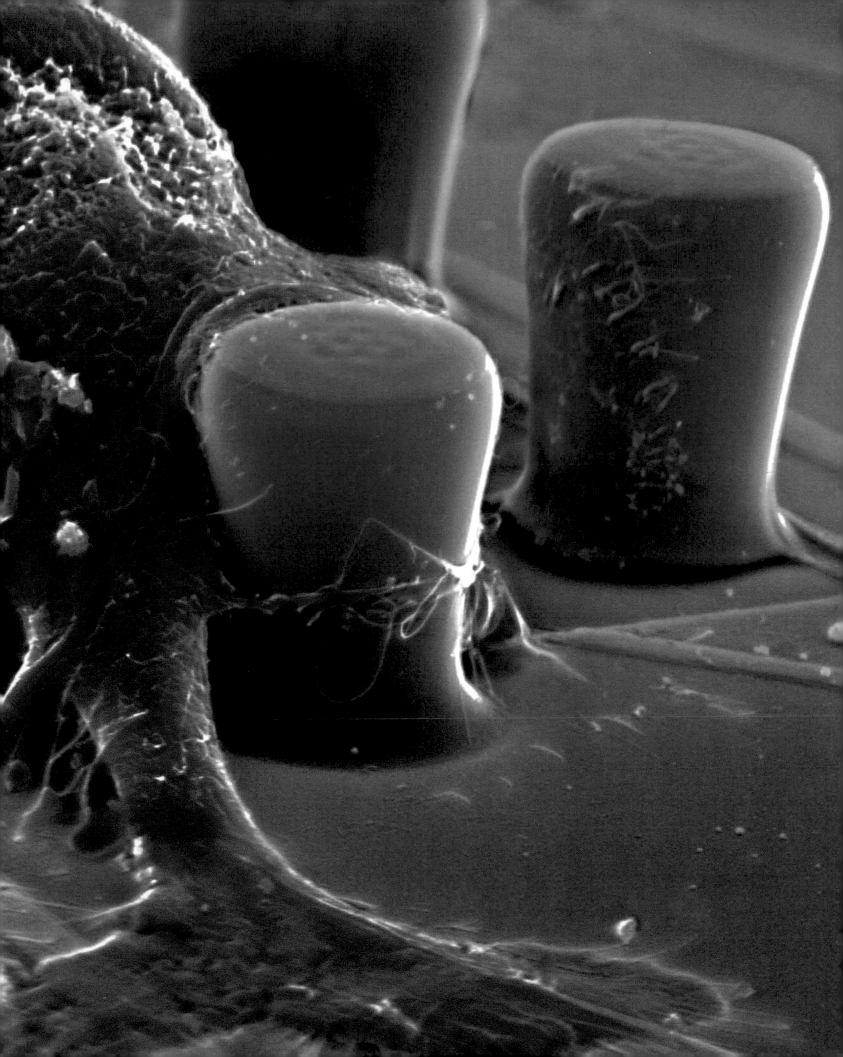

IV

Zoology

Sponge spicules

SPONGES ARE AQUATIC ANIMALS which filter food from the water. They consist of a loose aggregation of cells supported by a skeleton of spicules. Spicules are made of either calcium carbonate or silica in the shape of needles, stars, branching forms, commonly spine-like or hooked. The body of the sponge Tethya sp. is supported by spicules of silica along with fibers of the protein spongin. The shape of spicules is unique to each species of sponge and can be used as an identification tool.

SPICULES FROM THE SPONGE TETHYA MINUTA

32000✕

Rotifer colony

ALLIED TO ROUNDWORMS, this colony is composed of between 50–100 trumpet-shaped individuals attached at the base of their feet, with their bodies radiating outward from a common center. Hair-like cilia on the head (white) of each worm catch food particles suspended in the water and pass them to the mouth. The colony is a free-swimming plankton in freshwater: coordinated beating of the cilia propels it through the water.

LIGHT MICROGRAPH OF A COLONY OF ROTIFER WORMS, CONOCHILUS HIPPOCREPIS

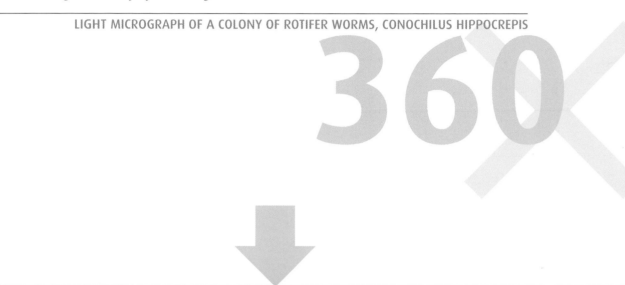

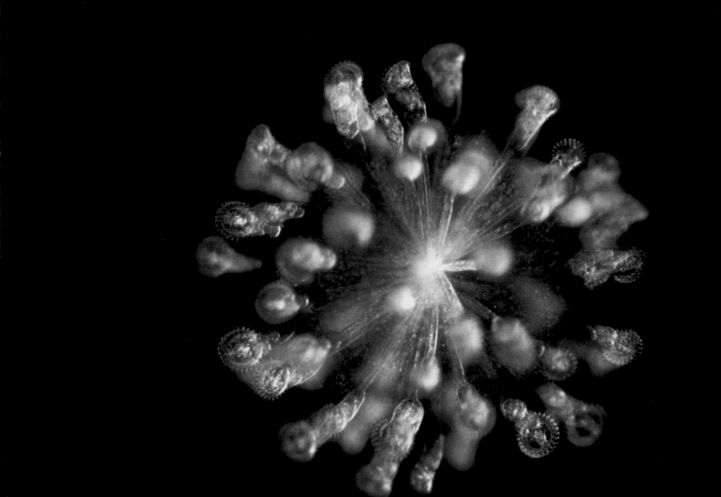

Butterfly eggs

BUTTERFLIES OFTEN LAY their eggs on plants, thus providing the newly-hatched larvae with a ready food supply. These eggs have already hatched, the new larvae (caterpillars) having bitten their way out by making a circular opening, leaving a lid on many of the eggs.

CLUTCH OF UNIDENTIFIED BUTTERFLY EGGS ON A RASPBERRY PLANT

155

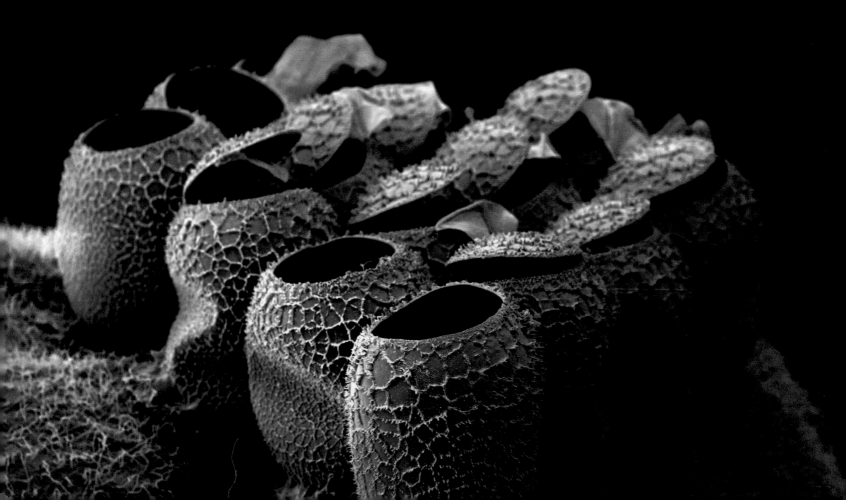

Caterpillar feet

THE RINGS OF TINY HOOKS on the end of each foot allow the caterpillar (larva) to maintain a firm hold on its substrate, usually a leaf, as it eats. Long touch-sensitive hairs are visible. The flexible and wrinkled body of the worm enables it to grow rapidly in size.

PAIRED FEET ON THE UNDERSIDE OF AN UNIDENTIFIED CATERPILLAR (ORDER LEPIDOPTERA)

60X

Tarantula eyes

TARANTULAS ARE VENOMOUS spiders that hunt insects, lizards and birds, requiring good sight to capture their prey with a powerful bite. The Mexican red-kneed tarantula, however, does not have big eyes nor particularly good vision, relying more on its sensitivity to noise and vibrations, detected by hairs on its legs. The eyes are located on a raised bump on the head. This type of tarantula has a body length of around 4 inches (10 cm). It is nocturnal (active at night), found in the deserts of Mexico and the southern USA.

THE EIGHT EYES (TWO GROUPS OF FOUR) ON THE HEAD OF A MEXICAN
RED-KNEED TARANTULA (BRACHYPELMA SMITHI)

Jumping spider eyes

THIS SMALL TROPICAL SPIDER has strong hind legs and can jump up to 50 times its own body length. It hunts its prey rather than catching it in a web. Its excellent vision allows it to spot the prey from a distance. It then creeps up and pounces by jumping, whilst trailing a safety line of silk behind it.

FIVE OF THE EIGHT SIMPLE EYES (OCELLI, PINK) ON THE HEAD OF A JUMPING SPIDER (FAMILY SALTICIDAE)

80X

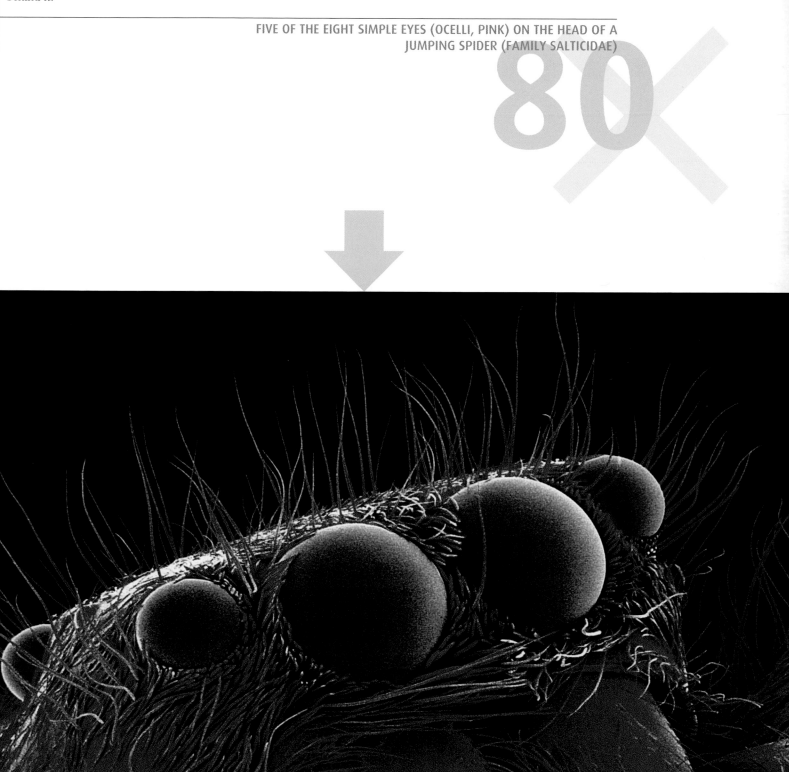

Household dust

DUST CONTAINS AN ASSORTMENT of substances which can cause asthma or other allergic reactions. The allergens seen here include: long hairs such as cat fur, twisted synthetic and woollen fibers, serrated insect scales, a pollen grain, plant and insect remains. Asthma or hay fever are often a hypersensitive and irritant response to substances that, in most people, are harmless. Airborne allergens may come into contact with the respiratory airways, skin or surface of the eye.

A SAMPLE OF ORDINARY HOUSEHOLD DUST

370✕

Gall mites

THESE PESTS ARE SO-CALLED because they form abnormal localized swellings (galls) on the plant, or leaf outgrowths, as a result of parasitic attack. They infest a wide range of plants, causing economic damage to food crops and forests. There are many hundreds of gall mites species, found on a variety of plants.

SEVERAL GALL MITES (ERIOPHYES SP.)

1115

Mite on honeybee

MITES ARE RELATED TO TICKS AND MORE DISTANTLY TO SPIDERS, in having four pairs of legs. Honeybee mites feed on the body fluid of adult bees and their pupae. Female mites lay their eggs inside the wax brood cells that contain larvae. The mites can cause significant damage to a bee's development and infestation of a hive with mites can cause eventual destruction of the colony.

A MITE (VARROA SP., UPPER CENTER) ON THE THORAX OF A HONEYBEE (APIS MELLIFERA)

Bee mites

EVEN THE HUMBLE HONEYBEE has smaller creatures crawling around on it. Bee mites move through the body hair of the bee, holding on tightly, secure even when the bee is in flight. Some bee mite species feed by biting into the surface membranes of the bee, damaging it. In large numbers they can devastate colonies of bees, which is of particular concern since bees are often agriculturally very important: as both pollinators and honey producers.

MITES ON THE HAIRY BODY OF A BEE

50X

Male dust mite genital pore

DURING MATING, the male dust mite will mount the back of the female and rotate to face the opposite direction. He holds onto her by two paranal suckers (seen in the next image) located on the underside of his abdomen. Once in position, the penis becomes erect through hemolymphatic (body fluid) pressure and penetrates the female's genital opening. Thousands of these tiny dust mites live on the furniture and fabric of an average home.

GENITAL PORE OF A MALE DUST MITE (DERMATOPHAGOIDES SP.) THE TWO STRUCTURES (CENTER) ARE THE PARAGYNAL LIPS WHICH COVER THE PENIS

4325X

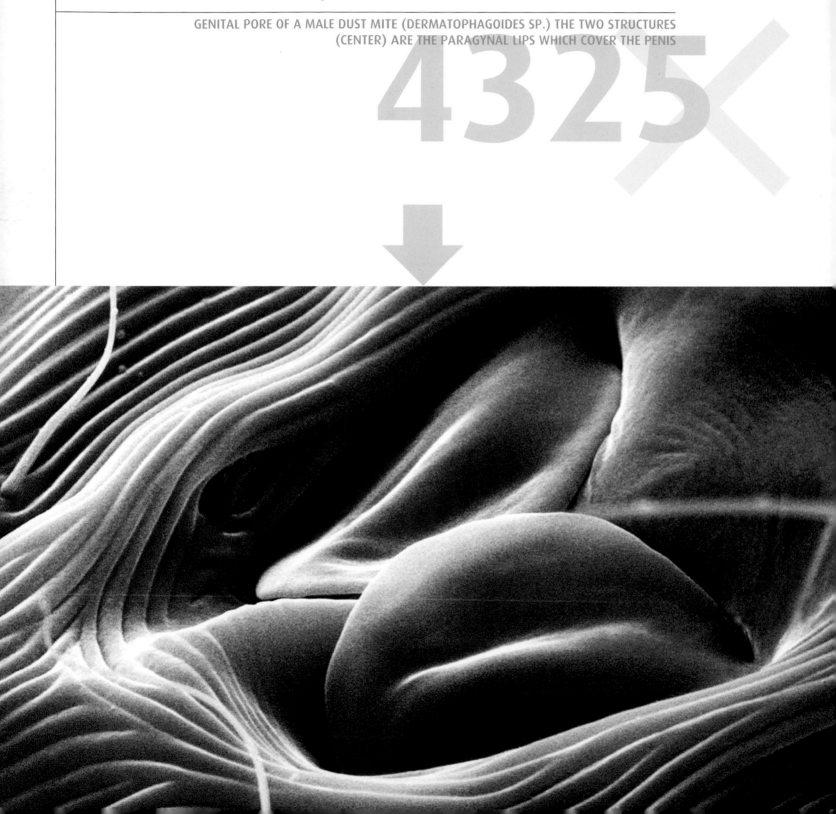

Dust mite suckers

BESIDE THE ANUS of a male dust mite are two rounded suckers on the underside of the abdomen. During mating, the male mounts the back of the female and holds onto her using these two specialized suckers whilst copulation occurs. Dust mites are found in household dust and although they are too small to be seen by the naked eye, they are related to spiders and scorpions. They eat the dead scales of human skin found in dust.

PARANAL SUCKERS OF A MALE DUST MITE (DERMATOPHAGOIDES SP.)

2200X

Mating schistosome flukes

As adults, these trematode worms which cause the disease of bilharzia (schistosomiasis) in humans, enter the human body and live in the veins around the large intestine and bladder, attaching themselves to the wall by a sucker on their head. During mating, the larger male holds the female in a groove which he forms by folding the sides of his body around her. The female produces many eggs. This serious parasitic disease affects 200 million people worldwide.

PAIR OF BILHARZIA WORMS (SCHISTOSOMA MANSONI) DURING COPULATION

173✕

Hawkmoth tongue

THE PROBOSCIS is an elongated mouthpart used by moths and butterflies
to suck up nectar and other liquids. Although it appears split down
the middle, the proboscis is held together to form a tube and may be
extremely long, often longer than the body of the moth. When not in use
it is coiled into a spiral. The Hummingbird Hawkmoth hovers in front
of flowers while drinking nectar using this long tongue. It lives in Europe,
and is active during the day.

END OF THE TONGUE (PROBOSCIS) OF A HUMMINGBIRD HAWKMOTH
(MACROGLOSSUM STELLATARUM)

1150X

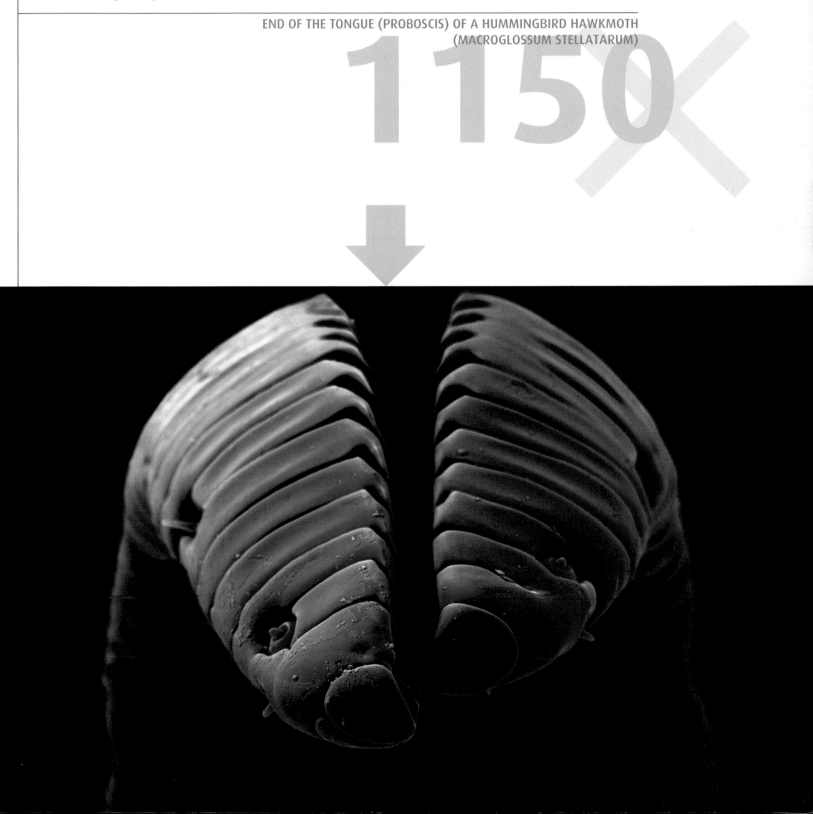

Blood-filled tick

So **FULL OF BLOOD** is this tick that the legs are protruding straight from its body on either side of its mouthparts (dark brown). This is a common sheep tick (Ixodes ricinus), the principal vector of Lyme disease in Europe. It is common in the damp underbrush of European woods and attacks various domestic and wild animals, including dogs and humans. It carries the bacterium (Borrelia burgdorferi) that causes Lyme disease.

THE TICK IS SWOLLEN IN SIZE AFTER FEEDING ON THE BLOOD OF ITS MAMMAL HOST

40X

Aphid

APHIDS ARE FOUND IN GROUPS, such as this one seen standing on top of others, infesting plants in great numbers. An aphid's mouthparts, a cutting and sucking instrument, allows it to penetrate the plant surface in search of sugary substances. In this way they cause the leaves of a plant to wilt and fall. Aphids are a major economic pest, transmitting virus diseases to the plant while feeding. The green peach aphid feeds on crops such as the potato, causing significant damage.

GREEN PEACH APHID, MYZUS PERSICAE

190X

Blowfly laying eggs

BLOWFLIES LAY THEIR EGGS on dead bodies, a behavior studied by forensic scientists. A blowfly detects a dead body by the odor of decomposition, and can arrive at a corpse minutes after death and lay up to 300 eggs. The decaying flesh is food for the maggots (blowfly larvae) that hatch from the eggs within 24 hours. These eggs are laid in patches around moist orifices such as the nose, ears and eyes, as well as open wounds. Fresh and unhatched blowfly eggs will indicate a very recent time of death.

A FEMALE BLOWFLY (LUCILIA SP.) LAYING HER EGGS

16

Head louse

THIS PARASITE INHABITS the hair of the head, gripping the hair using its well developed six legs, each of which ends in a curved claw. It glues its eggs (called nits) to the shafts of individual hairs. It feeds on the host's blood, often causing intense itching and irritation of the scalp. A head louse infestation may be treated with medicated shampoos.

A HUMAN HEAD LOUSE (PEDICULUS HUMANUS CAPITIS) CLINGING TO A HAIR

82X

Mosquito head

THE TWO LARGE COMPOUND EYES dominate the head. Each eye is composed of multiple facets, called ommatidia. Eyes of this design allow for the detection of small movements, but are weak in distinguishing detail. The mosquito also has two antennae, the bases of which are seen between the two eyes. Many hairs are also present around the head. Several mosquito species are vectors for human diseases such as malaria and yellow fever.

HEAD OF A MOSQUITO (FAMILY CULICIDAE) SHOWING THE EYES AND ANTENNAE

370

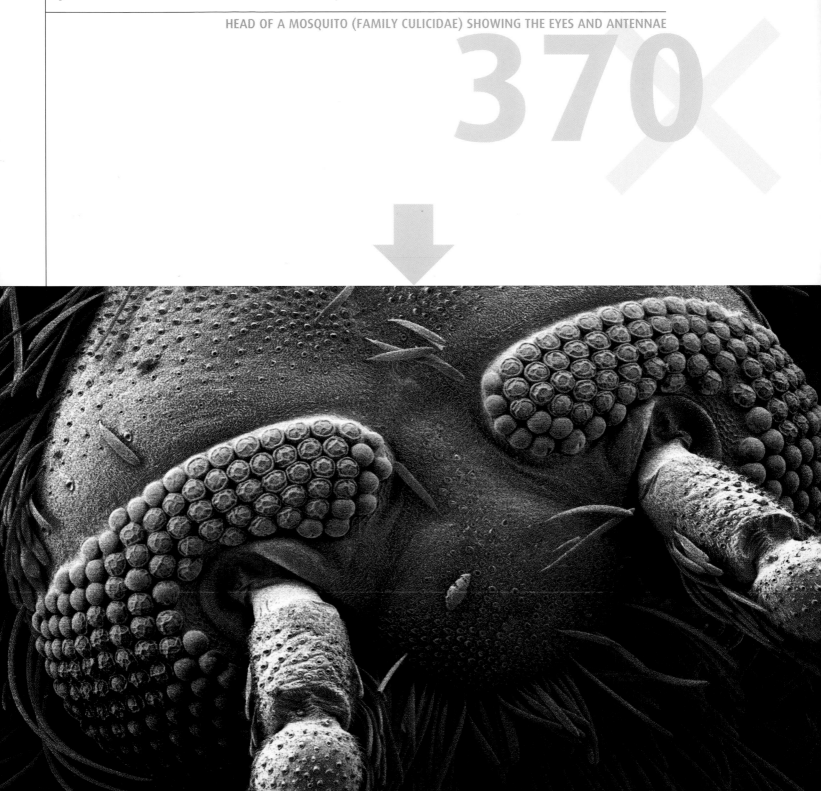

Wasp head

TWO LARGE COMPOUND EYES (purple) are seen on either side of its head. Coiled antennae, used to sense odors, sprout from between the eyes. The wasp's powerful biting jaws (mandibles) are just below the eyes. On the first thoracic segment are the wasp's front legs. Wasps are vital for natural biocontrol of insect pests. Nearly every pest insect species has a wasp species that is its predator or parasite.

FRONT VIEW OF A WASP'S HEAD (ORDER HYMENOPTERA)

30X

Moth antenna

PROTRUDING FROM THE HEAD, a moth's two long antennae are its major sense organ. As on the rest of the moth's body, each antenna may be covered in flattened scales (as seen here) as well as hairs. The sensory hairs are used for touch, detecting movement and vibrations. They act like a nose to sniff out food sources. And they also locate sex pheromones, the odors which female moths emit to attract the male during the mating season.

THE ANTENNA OF A MOTH (ORDER LEPIDOPTERA)

460✗

Butterfly scales

THESE BODY SCALES HAVE an intricate design and overlap like the tiles on the roof of a building. They allow heat and light to enter, and also insulate the insect. They may also be highly colored to form an identification and signaling system for the butterfly. The metallic appearance of the scales is due to ridges along their length.

SCALES FROM THE WING OF A PEACOCK BUTTERFLY (INACHIS IO)

420✕

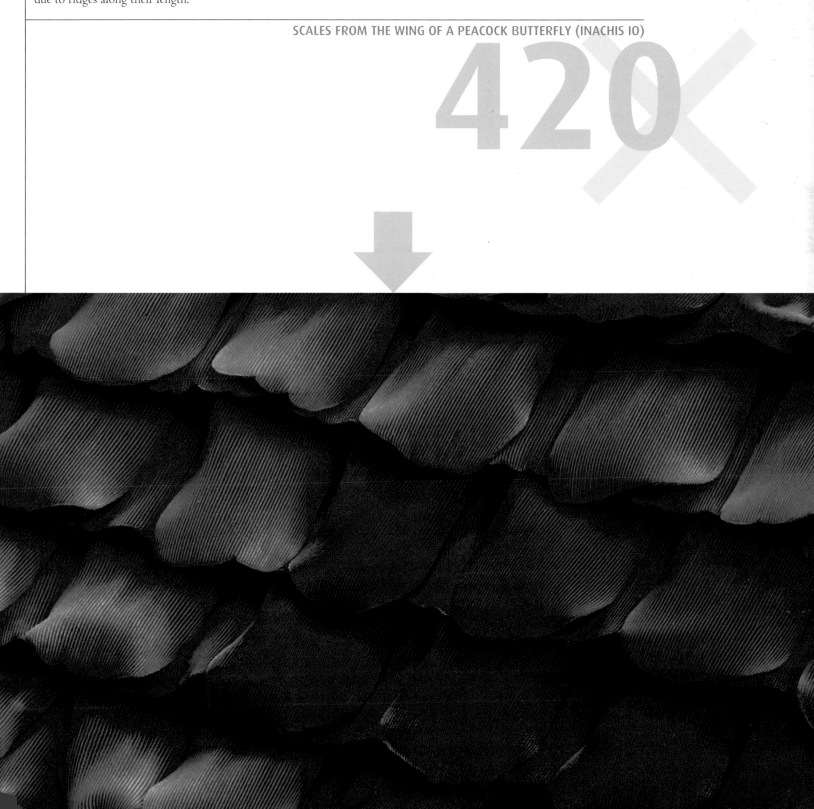

Butterfly scales

THE COLOR OF THE BUTTERFLY'S WING is determined by the pigmentation of the scales and also by diffraction of light by the surface of each scale. The scales grow out of the epidermal membrane. The swallowtail butterfly inhabits fields and meadows in Europe, North Africa and parts of Asia.

SCALES ON THE WING OF THE SWALLOWTAIL BUTTERFLY (PAPILIO MACHAON)

585

Shark skin

THESE SHARPLY POINTED PLACOID SCALES are also known as dermal teeth or denticles. They give the shark's skin the feel of sandpaper. The tip of each scale is made of dentine overlayed with dental enamel. The lower part of each scale, which anchors it into the skin, is made of bone. The scales disrupt turbulence over the skin, considerably reducing the drag on the shark as it swims. This design has been investigated by engineers for use on the surfaces of aircraft and boats.

SCALES ON THE SKIN OF A SHARK

135×

Snake skin

A SNAKE'S SKIN, contrary to the myth of it being slimy, is dry and mostly smooth. The scales (oval, green/blue) provide color and camouflage markings, protection, and can assist the snake to grip the ground when it is moving. Skin between the scales gives flexibility, but to grow a snake must shed its skin several times a year. The pattern of scales, type and configuration in snake skin is as unique as a fingerprint.

SHED SKIN OF AN UNIDENTIFIED SNAKE

64

Iguana tail spines

THE GREEN IGUANA, a large type of lizard, is native to tropical South America, but makes a popular pet. Their tough skin consists of scales of keratin and running down the tail dorsally from high up its back is a row of protective spines (seen here). Iguanas, like other lizards and snakes, shed their skin periodically as they grow and new scales grow underneath the old ones. The green iguana lives in tropical forests and enjoys climbing trees, assisted by long claws.

SKIN FROM THE TAIL OF A GREEN IGUANA (IGUANA IGUANA) SHOWING SCALES AND SPINES

160✕

Silkworm proleg

CATERPILLARS GENERALLY HAVE THREE PAIRS of jointed legs behind the head, and a number of fleshy prolegs further along the body. At the end of the proleg is a double row of curved hooks, which help the caterpillar grip the surface when walking. The caterpillar's body is covered in fine hairs which are touch sensitive. The silkworm produces threads of silk with which it constructs a cocoon. Silkworms are farmed for their silk, which is woven into cloth.

PROLEG OF A SILKWORM MOTH CATERPILLAR (BOMBYX MORI)

184

Vine weevil foot

STANDING ON THE SURFACE OF A LEAF, the foot of this insect, a type of beetle, comprises a pair of adhesive hairy pads (lower center) and a pair of hooked claws (center right). The pads are used for walking on smooth surfaces, while the claws are used on surfaces that they can grip. This beetle cannot fly but is a very active walker. It is a pest that attacks a wide range of plants including the yew, rhododendron, azalea, laurel, juniper, grape, holly and strawberry. All the adults are female and reproduce by parthenogenesis (asexual reproduction).

CLAWED FOOT OF THE BLACK VINE WEEVIL, OTIORHYNCHUS SULCATUS

140✕

Gecko foot

THE FOOT IS COVERED with ridges and microscopic hairs, which enables the
gecko to cling to very smooth surfaces such as windows and ceilings.
Geckos are slender nocturnal lizards. They feed on insects and are found
in a wide range of habitats, from deserts to rainforests. They are also
often seen climbing the walls of houses in warm climates.

UNDERSIDE OF A GECKO'S FOOT (FAMILY GEKKONIDAE)

95X

Feather

FEATHERS ARE MADE OF KERATIN and serve a variety of functions: for decoration and display by the bird, for insulation against the cold and wet, as well as for flight. A down feather is smaller and more fluffy in appearance than a flight feather. Located deeper on the body of the bird, particularly on its torso, it provides warmth and when these feathers are shed they are often chosen to line the nest. Goose down is also popularly used in duvets and sleeping bags.

STRUCTURE OF A GOOSE DOWN FEATHER AT LOW MAGNIFICATION

70X

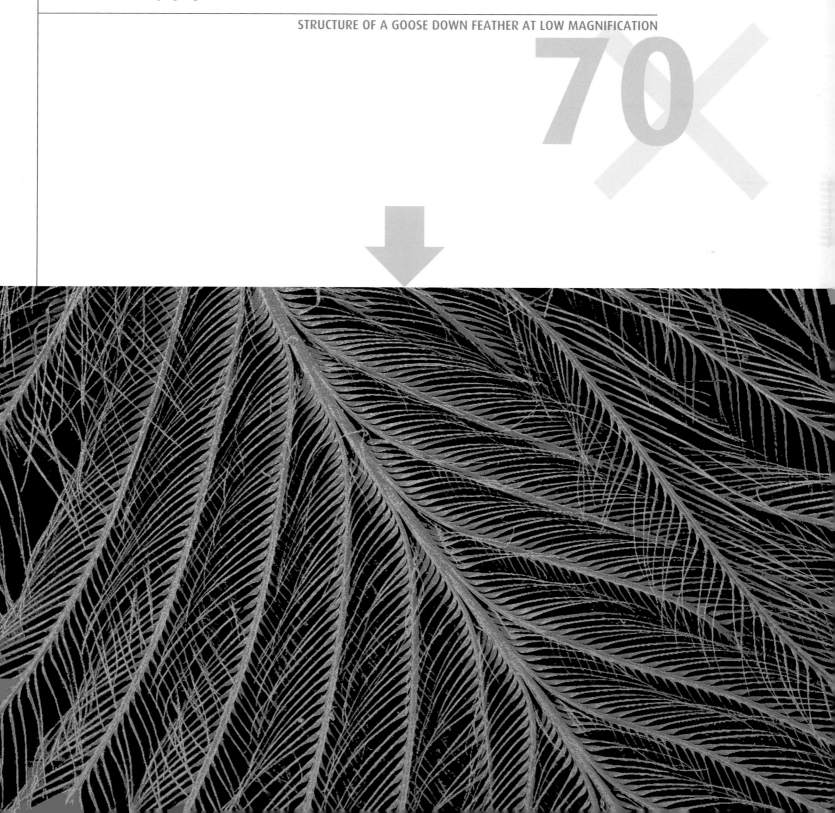

Feather

THE FLAT, BROAD VANE of a feather is formed from comb-like rows of
filaments called barbs, which project from either side of a central shaft
(rachis, diagonally from bottom left to top right). Each barb has rows of
minute filaments called barbules (orange). The barbules on one side bear
hooks, while those on the other have a groove. This arrangement hooks
adjacent barbs together, linking the whole structure, making it flexible
and light, which is ideal for flight. Goose down is the insulating layer of
feathers close to the skin that helps to keep a goose warm and dry.

STRUCTURE OF A GOOSE DOWN FEATHER AT HIGH MAGNIFICATION

1400×

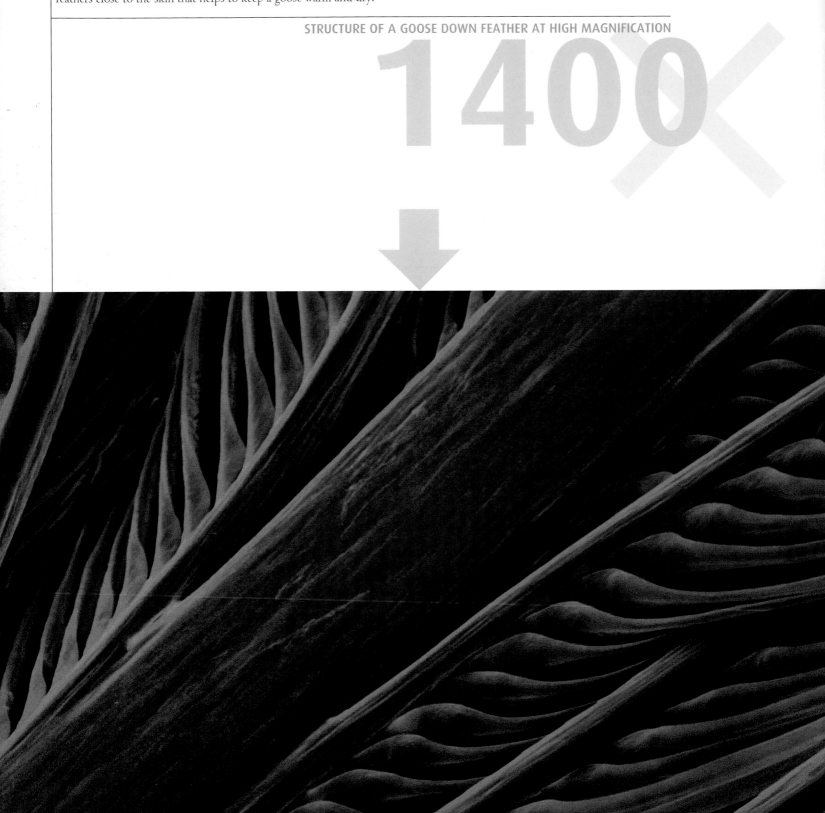

Bird bone

THE COMPLICATED STRUCTURE of a jaw joint includes spaces for the attachment of muscles, tendons and ligaments. This allows the bones of the joint to move independently (articulate) while remaining connected. In birds, the large jaw bones have been reduced in size and covered by keratin (like in an antler or fingernail) to form the beak.

PART OF THE JAW SOCKET FROM A BIRD SKULL

75×

V

Minerals

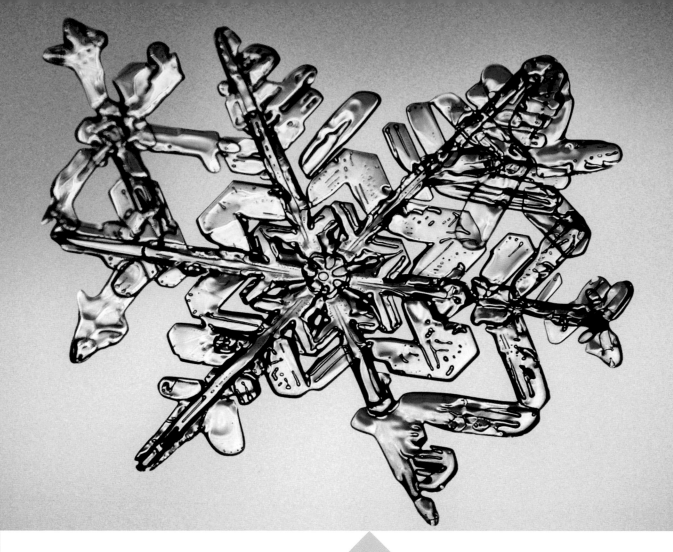

Dendritic snowflake

SNOWFLAKES ARE SYMMETRICAL ICE crystals that form when air has a temperature near the freezing point of water. If the air is calm then a symmetrical, hexagonal (six-sided) snowflake can form, as seen here and in the next image. When growth is fast and unstable, branching patterns create a dendritic snowflake, this being a typical example. Slower growth in calm air allows the straight edges of a plate snowflake to form, seen on the next page. No two snowflakes are the same, as each experiences a wide range of conditions as it forms inside a cloud.

LIGHT MICROGRAPH OF A DENDRITIC (BRANCHING) SNOWFLAKE

57

Plate snowflake

THE TWO MAIN GROWTH PATTERNS observed when snowflakes form are
faceting and branching. When growth in the air is slow, the straight
edges, lines and hexagonal shapes of faceting create a plate snowflake.
As a snowflake grows, instabilities cause branching and the star-shaped
patterns of a dendritic snowflake. The type of growth also depends on
temperature, which changes in a different way for each snowflake,
causing the wide variation of patterns in snowflakes.

LIGHT MICROGRAPH OF A HEXAGONAL PLATE SNOWFLAKE

104

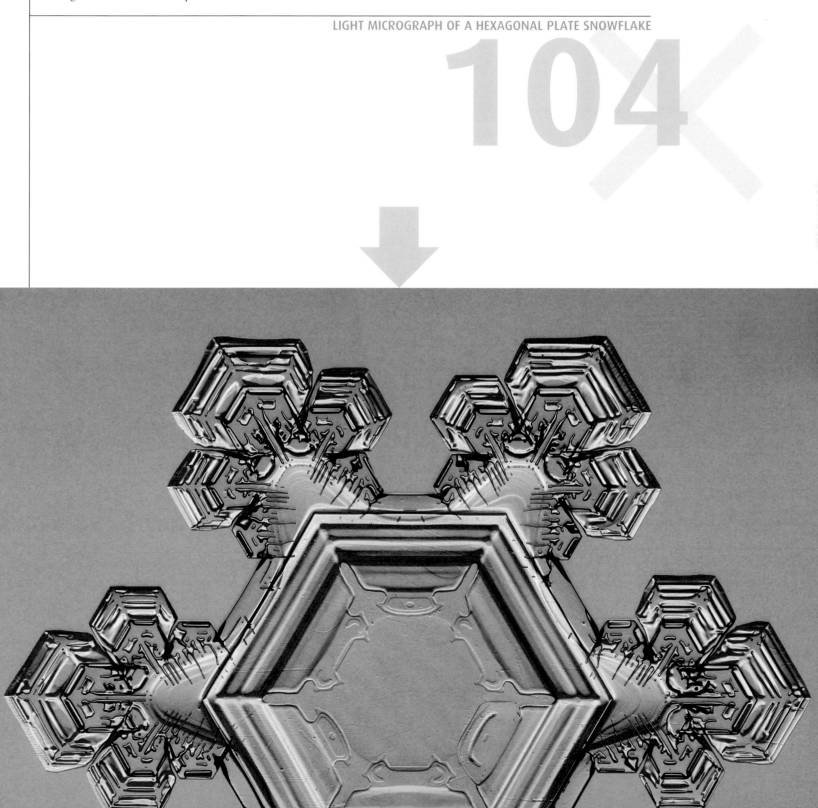

Calcium phosphate crystal

CRYSTALLINE MATERIALS have their atoms arranged in regular lattices that can form geometric shapes like this. Crystals of calcium phosphate contain both calcium (Ca) and phosphate (PO4) ions, but the precise ratio varies according to the crystal structure. Calcium phosphates are used in dental materials and also in foodstuffs.

GEOMETRIC STAR-SHAPE OF A CRYSTAL OF CALCIUM PHOSPHATE

16000✕

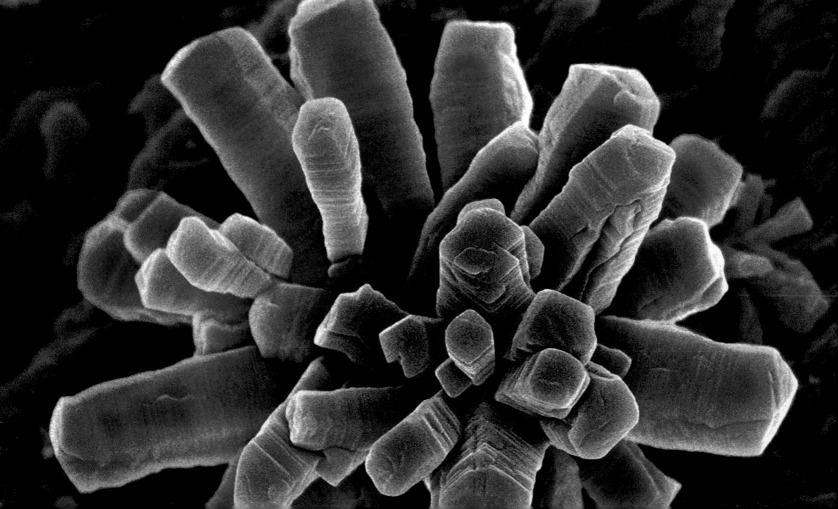

Calcium crystals in compost

CALCIUM IS A SOFT, silvery-white metal. It is extremely common in nature, a constituent of body fluids, cells, bones and teeth. In a compost heap of vegetable trimmings, grass cuttings and other degrading plant material, calcium will leach out. Calcium contained in the decomposing layers of material is subject to great changes in acidity (pH). When the pH is low (acid) calcium exists in a creamy-white solution. When the pH rises (alkaline) the calcium crystallizes, as seen here.

NATURAL CRYSTALS OF CALCIUM IN A COMPOST HEAP

1340✕

Folic acid

FOLIC ACID IS A COENZYME needed for forming body protein and hemoglobin in red blood cells. It is an important vitamin which has been shown to reduce the risk of pregnant women giving birth to children with spine or brain defects. These defects (such as spina bifida) can occur in the embryo in the first stages of pregnancy; women of childbearing age or in early pregnancy are encouraged to include folic acid in their diets. Dietary sources of folic acid include liver, leafy green vegetables, legumes like peas and beans, nuts, whole grains and brewer's yeast.

POLARIZED LIGHT MICROGRAPH OF CRYSTALS OF FOLIC ACID (FOLACIN), A MEMBER OF THE VITAMIN B COMPLEX

140

Vitamin C

VITAMIN C IS A WATER SOLUBLE VITAMIN which plays an essential role in the activity of many enzymes within the human body. It is necessary for the growth and maintenance of healthy bones, gums, teeth, ligaments and blood vessels. It is also involved in the production of neurotransmitters and in the stimulation of the immune system against infections. The main dietary sources of vitamin C are citrus fruits and green vegetables. Deficiency of vitamin C leads to scurvy, with swollen bleeding gums and anemia.

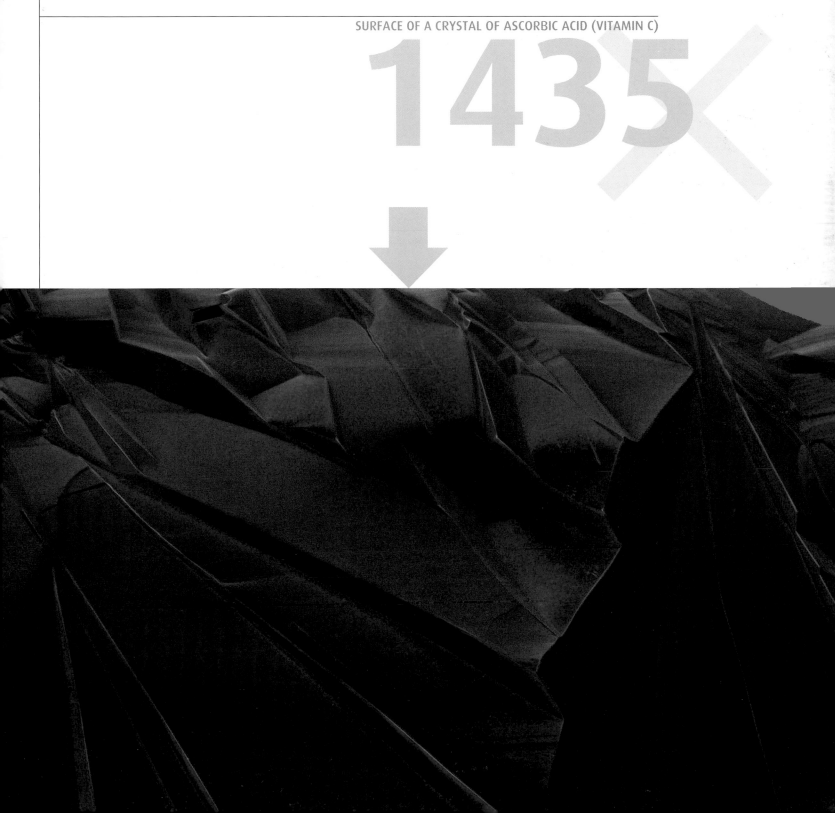

SURFACE OF A CRYSTAL OF ASCORBIC ACID (VITAMIN C)

1435×

Vitamin E crystals

VITAMIN E IS PRINCIPALLY concerned with protecting fats from oxidation in the body. It is needed for maintenance of normal cell structure, for sustaining the action of certain enzymes, and it protects the lungs and other tissues from damaging pollutants. Vitamin E is also believed to slow the aging of cells, and for this reason is often present in cosmetic products. The principal dietary sources of vitamin E are vegetable oils, nuts, meat, green leafy vegetables, cereals, wheat-germ and egg yolk. Because vitamin E is found widely in foods and stays in the body a long time, deficiency is rare.

POLARIZED LIGHT MICROGRAPH OF CRYSTALS OF ALPHA-TOCOPHEROL, THE MOST EFFECTIVE OF THE VITAMIN E GROUP

30X

Mescaline

MESCALINE IS PRODUCED from the dried tops, or buttons, of the Peyote cactus (Lophophora williamsii). The cactus is found growing wild in Texas and Mexico and was used extensively by both Native Americans and Aztecs in religious ceremonies. It is usually taken by chewing the buttons or making them into an infusion. Mescaline produces feelings of euphoria and an altered perception of time, space and color.

CRYSTALS OF MESCALINE, A HALLUCINOGENIC DRUG

890X

Morphine

MORPHINE IS USED TO TREAT SEVERE PAIN due to its potent analgesic (painkilling) properties. It acts on receptors that are naturally triggered by endorphins, the body's natural painkillers. Morphine is also abused as it produces feelings of euphoria and dreaminess. It is highly addictive and the body also develops tolerance quickly, so that larger doses are needed to achieve the same effect.

CRYSTALS OF MORPHINE, A NARCOTIC DRUG DERIVED FROM THE OPIUM POPPY

450✕

Palladium

THE CRYSTALS HAVE A CUBIC close-packed structure, with the crystal faces clearly seen. Palladium is the least noble of the platinum metals with a melting point of 1552 Celsius. It is used in alloys such as white gold. It serves as a catalyst, especially for hydrogenation. Palladium is also used in dental work and in electrical components.

CRYSTALS OF THE METAL PALLADIUM (SYMBOL: PD)

195X

Tungsten crystals

LIKE MANY METALS, tungsten exhibits a crystalline structure at a microscopic level. This gray, hard, metallic element is used extensively in steel alloys, where it imparts great hardness. These alloys are used in the manufacture of armor plate and cutting tools. Tungsten is also used in filament lightbulbs and thermionic valves due to its very high melting point (6192°F/3422°C).

CRYSTALS OF THE METAL TUNGSTEN (SYMBOL: W)

7250✕

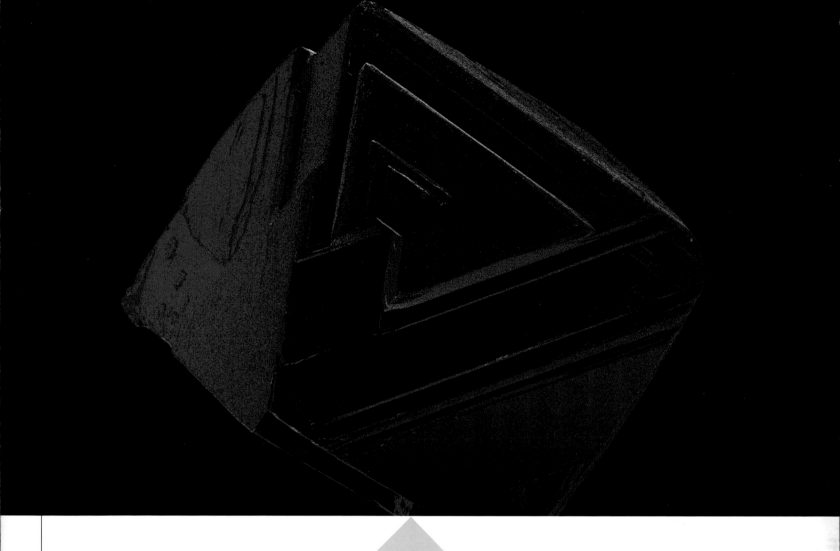

Microdiamond crystal

MICRODIAMONDS ARE DIAMONDS (pure crystals of carbon) which measure less than half a millimeter in any dimension. They are being studied as an alternative to the more expensive larger diamonds used in industry, for instance as abrasives or heat dissipators. Microdiamonds are thought to form either with larger diamonds in Earth's mantle (the layer beneath the crust), or in the ascending magma (molten rock) that can bring the larger diamonds to the surface. This example, the most common shape found, was discovered in Siberia, Russia.

AN OCTAHEDRAL (EIGHT-SIDED) MICRODIAMOND CRYSTAL

625X

Pesticide food contamination

FUNGICIDES ARE USED widely in agriculture to kill the fungi that damage food crops and other plants. The plants must be washed after harvesting, in order to reduce the contamination of human food with potentially toxic chemicals. Organic farming is a way to prevent this chemical contamination.

SINGLE CRYSTAL OF A FUNGICIDE ON A BROAD BEAN PLANT'S LEAF

7600✗

Sugar crystals

THESE CRYSTALS OF SUCROSE are also known as cane or beet sugar. They are a sweet, soluble crystalline carbohydrate. Easily digested, sucrose sugar forms a major source of energy in our diet. It is both a sweetener and preservative, and is used in cooking and the food industry. A high consumption is associated with obesity and tooth decay.

SEVERAL CRYSTALS OF SUCROSE, A NATURALLY-OCCURRING PLANT SUGAR

1150×

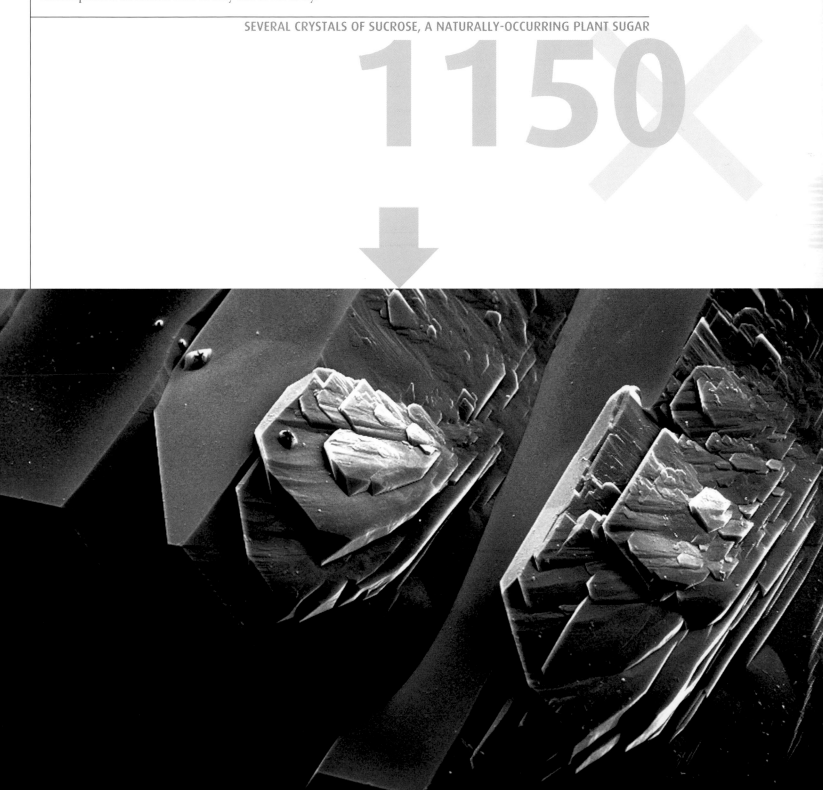

Sorbitol crystals

FIRST FOUND IN THE BERRIES of the mountain ash (Sorbus aucuparia), this is an alcohol derived from the sugar sorbose. It also occurs in other fruits, seaweeds and algae. For commercial use it is prepared from glucose. It is 60% as sweet as cane sugar, and 70% of sorbitol, once eaten, is converted to carbon dioxide without appearing as glucose in the blood. It is therefore widely used as a sweetener in diabetic confectionery.

APOLARIZED LIGHT MICROGRAPH OF CRYSTALS OF D-SORBITOL,
A DIABETIC SWEETENER

64

Kosher salt crystals

BECAUSE THE GRAINS of kosher salt consist of several crystals stuck together, they have a larger surface area than other types. This allows it to be used for absorbing liquids, such as in the curing of meats (including making meats kosher under Jewish food laws by removing blood traces - hence the name). Kosher salt also has a distinctive taste that is preferred by some cooks. It is obtained from salt mines.

GRAIN OF KOSHER SALT, A COARSE LARGE-GRAINED TYPE OF SALT
(A FORM OF THE MINERAL SODIUM CHLORIDE)

265X

Sea salt crystals

UNLIKE OTHER TYPES OF SALT, as seen on the previous and following pages, sea salt crystals contain traces of elements like zinc, magnesium, iron, calcium, potassium, manganese and iodine. This gives sea salts a distinctive taste. The size of the crystals can be changed by grinding to obtain fine-grained varieties from the coarse varieties.

GRAINS OF SEA SALT, A COARSE LARGE-GRAINED TYPE OF SALT FORMED NATURALLY BY THE EVAPORATION OF SEA WATER

265

Table salt crystals

TABLE SALT IS A FINE (small-grained) type of salt that is prepared by refining, grinding and recrystallizing natural salt. It also has additives, such as iodine and anti-caking agents to improve diet and make the salt grains flow more freely when added to food. Salt is obtained from salt mines or by evaporating sea water.

CRYSTALS OF COMMON TABLE SALT

495✕

Gallstone crystals

GALLSTONES FORM IN THE GALLBLADDER, a small organ that stores liquid bile produced by the liver until it is needed for the digestion of food. When there is an imbalance in the chemical composition of the bile, cholesterol and bile salts may crystallize, and over time these crystals can form into hard stones. Gallstones are usually symptomless unless one obstructs the bile duct, which can cause acute pain, jaundice and infection. Treatment is with drugs to dissolve the stones or ultrasound waves to break them up.

FRACTURED GALLSTONE, REVEALING ITS INTERNAL CRYSTALLINE STRUCTURE

920X

Silver-impregnated wound dressing

THE CHARCOAL AND SILVER IS THE BLACK LAYER, and the nylon fabric is the white layers. The silver has an antimicrobial action that kills bacteria in the wound. This dressing is also designed to absorb toxins and to reduce the smell of a wound. Actisorb Silver is used on malodorous, infected wounds, including fungal lesions, fecal fistulae, infected pressure sores, and leg ulcers. Actisorb Silver is manufactured by Johnson & Johnson Medical Ltd.

SAMPLE OF ACTISORB SILVER WOUND DRESSING. THIS CONSISTS OF AN ACTIVATED CHARCOAL CLOTH IMPREGNATED WITH SILVER

225X

Scales from a domestic kettle

KETTLE SCALES CONSIST of needles of calcium sulphate which precipitates
out of hard water into regular crystallographic shapes. The geologist
knows calcium sulphate as anhydrite. Its monoclinic crystal lattice and
flower-like clumps of needles are also the most common form in nature
and it is from rocks bearing anhydrite that water derives its hardness.
When water containing calcium sulphate is boiled or evaporates, the
calcium sulphate precipitates out of solution.

CRYSTALS OF CALCIUM SULPHATE FORMED INSIDE A KETTLE IN A HARD WATER AREA

760

Rusty nail surface

THE SURFACE APPEARS uneven due to patches of rust and oxidation and it is weakened from cracking. Rusting is an electrolytic process in which iron reacts with water and oxygen to form an hydrated iron oxide. Rust is more brittle, porous and bulky than iron, and is weaker.

CORRODED SURFACE OF A RUSTY METAL NAIL. CORROSION IS THE DESTRUCTION OF THE METAL BY CHEMICAL REACTIONS

2300✗

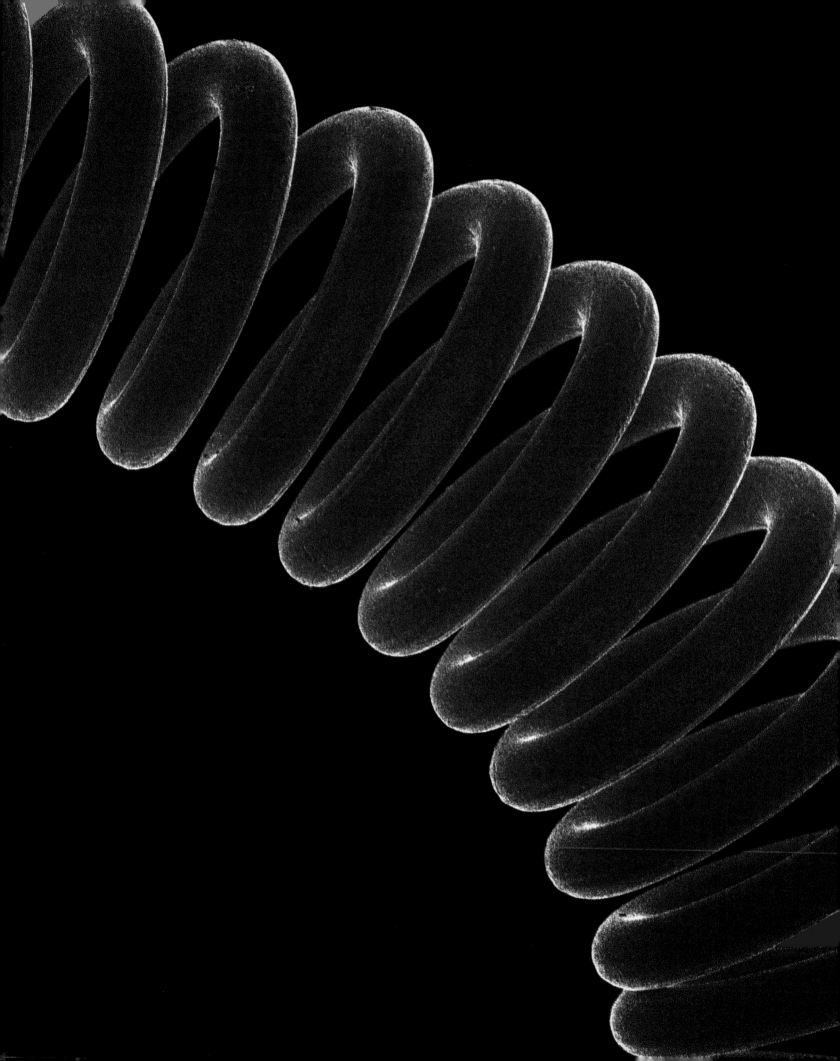

VI

Technology

Touch fastener

TOUCH FASTENER IS USED AS A COMMON FASTENER on clothes and shoes. It is a nylon material manufactured in two separate parts; one with a hooked surface (blue, at lower image) and the other with a smooth surface made up of a series of loops (green, at upper image). The loops are loosely woven strands in an otherwise tight weave. The hooks are loops woven into the fabric and then cut. When the two surfaces are brought together they form a strong bond, which can be peeled apart.

NYLON HOOKS AND LOOPS IN TOUCH FASTENER MATERIAL

60X

Wound dressing

THIS TYPE OF MEDICAL FIRST-AID dressing for minor cuts, abrasions and wounds has a waterproof and germ-proof outer layer (at top, brown), but allows the wound to breathe and is able to absorb any exudate from the wound. The absorbent layer is yellow, and the non-stick inner surface with pores is at bottom. The structure of this dressing creates an environment around the wound that hastens the healing time and minimizes scars.

POLYURETHANE (PU) FOAM WOUND DRESSING

45X

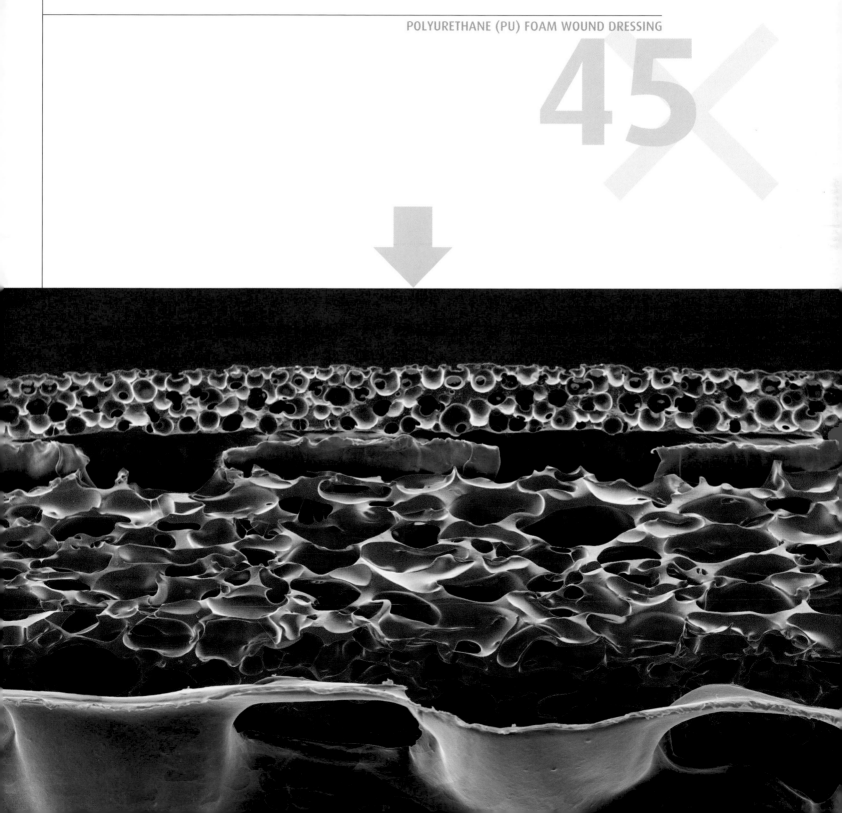

Ceramic superconductor

SUPERCONDUCTIVITY OCCURS IN CERTAIN materials at extremely low temperatures, with high-temperature superconductors working at the relatively higher temperatures of -193°F (-125°C). Mixing two or more metals that aren't superconducting can turn the alloy into a superconductor where electrical resistance vanishes and electrons flow freely. One application of superconductors is as the magnets in MRI body scanners. High-temperature ceramic superconductors were discovered in 1987 and although they are still the product of research, they have a range of potential applications in technology.

HIGH-TEMPERATURE CERAMIC SUPERCONDUCTING ALLOY. THIS METAL ALLOY IS MADE UP OF LANTHANUM, BARIUM AND COPPER-OXIDE

12860✕

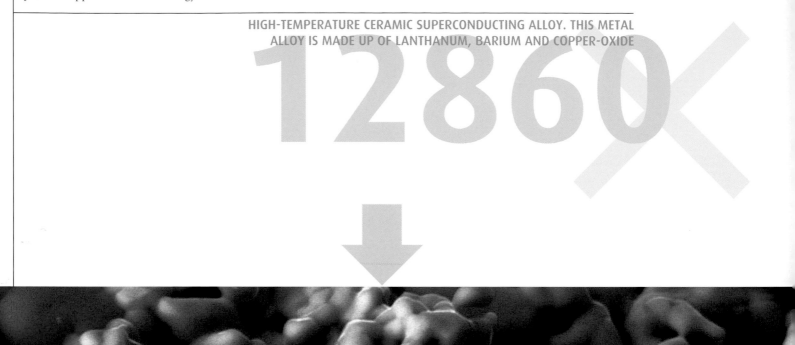

Tungsten oxide crystals

TUNGSTEN OXIDE (CHEMICAL FORMULA $W.O_3$) is the final product from heating the metal tungsten in excess oxygen. The crystals form a powder which is mainly used as a pigment in yellow glazes for ceramics. It can also be used in the manufacture of alloys and in the fireproofing of fabrics.

CLUSTER OF POINTED TUNGSTEN OXIDE CRYSTALS GROWN IN AIR

300X

Filter paper

CELLULOSE IS A POLYSACCHARIDE that is the main constituent of all plant tissues and fibers. Filter paper is a porous material used for separating the liquid and solid parts of a suspension and, being made from wood pulp, is almost pure cellulose.

CUT END OF A PIECE OF FILTER PAPER SHOWING ITS MANY CELLULOSE FIBERS

415×

Surface of a shoe insole

AT THIS MAGNIFICATION THE FIBERS of the woven top layer can be seen with its porous latex binder filling the gaps between the weave. The fibers provide the strength and body of the material, while the rubbery foam adds support and comfort, and its porous nature allows the foot to breathe.

SURFACE OF A FOAM/FIBER SHOE INSOLE

115✕

Sweat absorbent material

THESE MAN-MADE FIBERS LOOSELY arranged are covered with different coatings on the outside and inside. These prevent the cloth becoming wet during wear, by absorbing the moisture and then directing it away to the outside.

FIBERS OF A SWEAT ABSORBENT FABRIC

3830

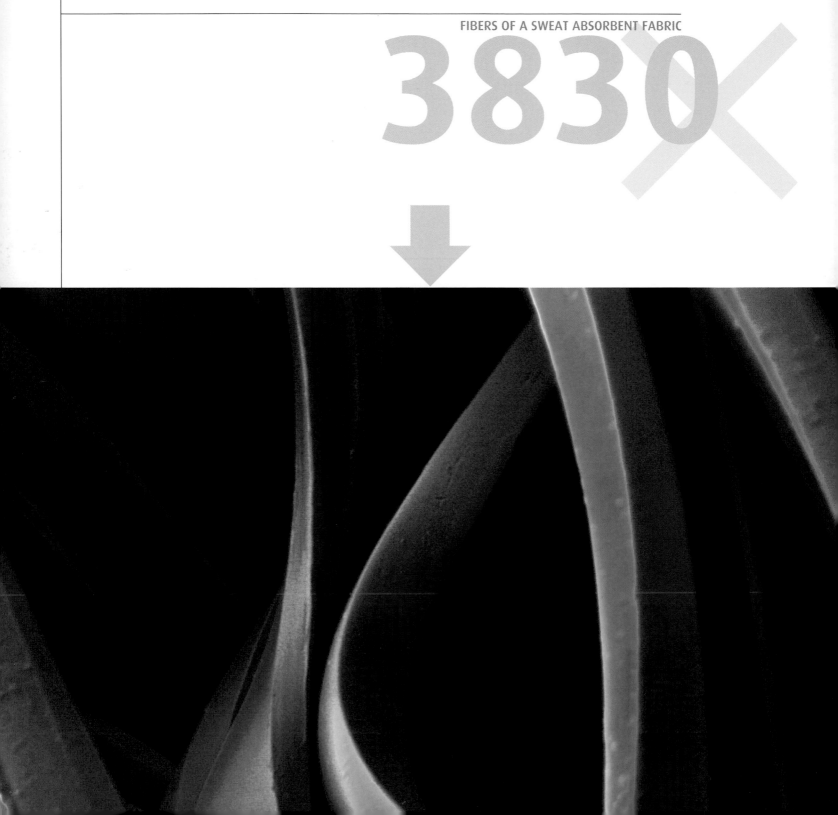

Nylon stocking

NYLON IS A POLYAMIDE SUBSTANCE, and the first synthetic fiber ever to be produced. Nylon fibers are stronger and more elastic than silk and are relatively insensitive to moisture and to fungal mildew. Nylon is used in the manufacture of textiles (particularly hosiery and woven goods, including carpets), molded articles and medical sutures.

THE LOOSE WEAVE OF A NYLON STOCKING (WOMEN'S TIGHTS)

295

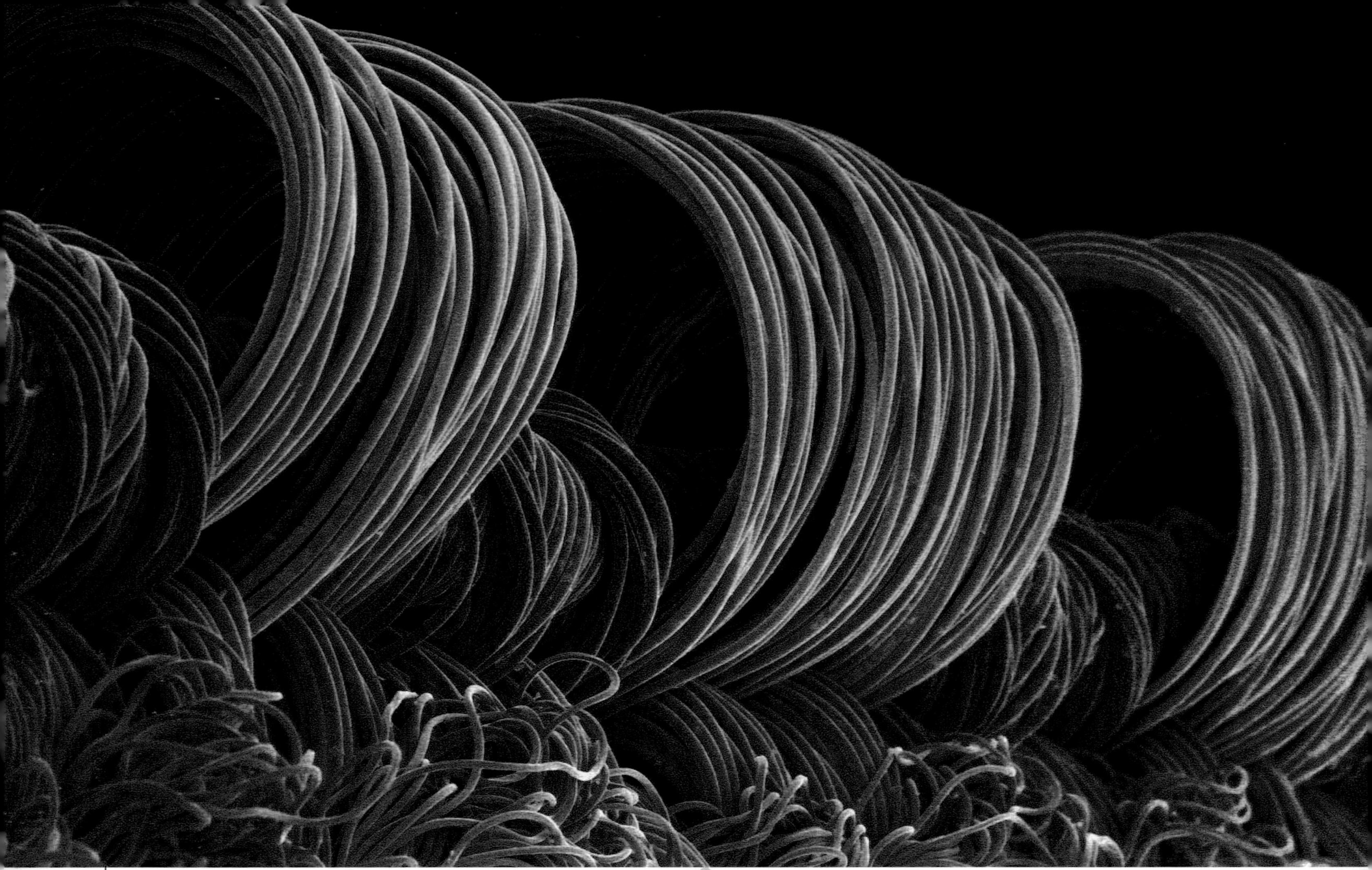

Fibers in a bra

THE COMPLEXITY OF STITCHING which can be achieved by a textile machine at the edge of a fabric is well illustrated here. The fabric of the bra (yellow, at bottom) has been edged with orange fibers to prevent fraying. While looped fibers (at top) help to reduce the itchiness and discomfort on the skin of this tight-fitting material.

LOOPED SYNTHETIC FIBERS ALONG THE EDGE OF A BRA

310

Waterproof clothing

WATERPROOF FABRICS CAN WITHSTAND a water pressure of over 1000 millimeters without leaking while allowing water vapor to penetrate. This fabric of outdoor clothing is made of cotton fibers tightly woven to enable the fabric to resist damage by water. The outer surface is covered with a polyurethane layer with many tiny pores. These allow humid air and water vapor to come out of the garment, but the small pore size prevents liquid water from entering due to surface tension. This type of coating is known as breathable waterproofing.

FABRIC FROM A PIECE OF WATERPROOF CLOTHING WITH WATER DROPLETS ON IT

90✕

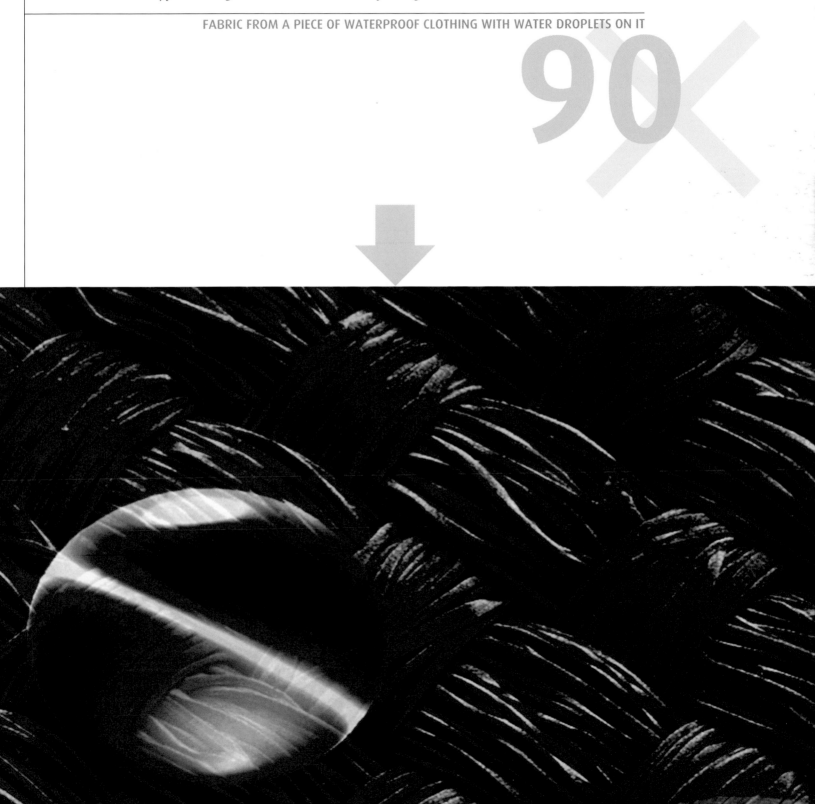

Raincoat material

THE FABRIC, FORMED BY a tight cross-weave of fibers, is covered with a water-repellant coating which holds rain drops outside. It also contains pores which allow condensation formed by the body's own heat to escape from the garment.

CROSS-SECTION THROUGH MATERIAL USED TO MAKE RAINCOATS

514

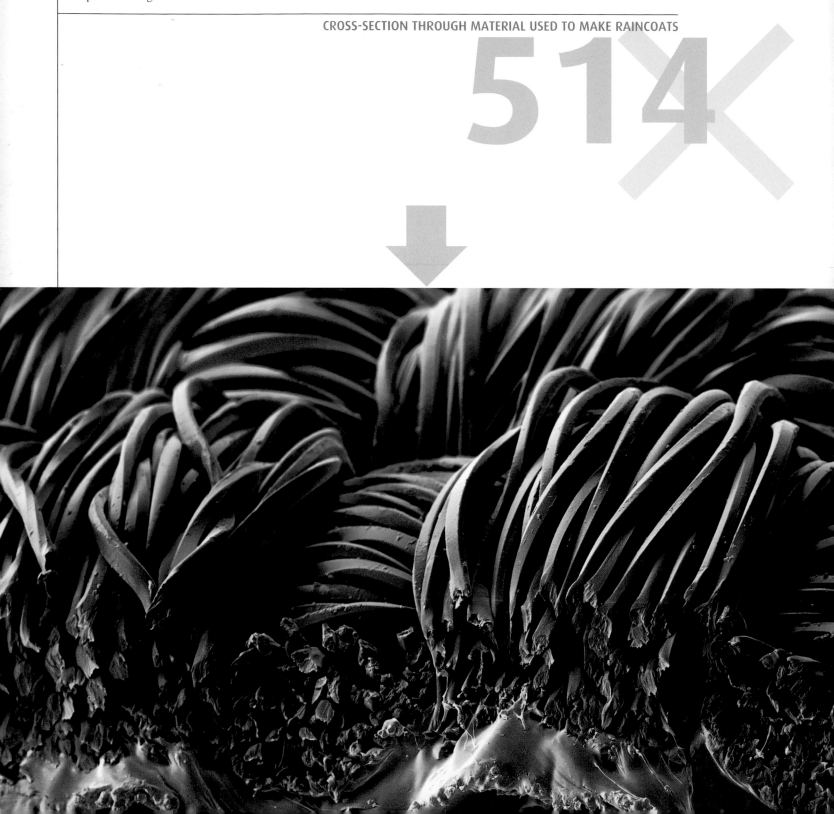

Toilet paper

THIS SOFT TOILET PAPER is made from pure bleached chemical woodpulp
without further additions—no sizing (sealing on the paper surface) or
alum. It is produced on large, fast-running paper machines. Although
vegetable fibers and rags are still used, modern paper is predominantly
made from wood, in particular spruce, hemlock, poplar, pine and fir.

PAPER FIBERS (CELLULOSE) THAT MAKE UP TOILET PAPER

190X

Cigarette paper

THE CRYSTALS (BLUE) ARE additives that keep the lit cigarette burning by
producing oxygen. The cellulose fibers making up the paper are also
seen (strands).

CLOSE-UP OF THE SURFACE OF CIGARETTE PAPER

650X

Dental crown

If the top of a tooth is badly cracked, decayed or broken, it can be replaced by an artificial crown. The remaining undamaged part of the tooth is cut down to a peg to receive the hollow crown. This crown is then shaped to form a replica of the original tooth and cemented in place. Here, the crack between the tooth and crown allows bacteria to enter and cause secondary tooth decay.

JOIN BETWEEN A TOOTH (LOWER HALF OF IMAGE) AND A CERAMIC DENTAL CROWN (BLUE, UPPER FRAME). DEBRIS HAS BECOME ATTACHED TO THE TOOTH WHEN IT WAS EXTRACTED FOR THIS IMAGE

160X

Surgical thread

The thread is only 50 microns thick (0.05 millimeters). This is less than the thickness of a human hair. Such threads are made from microfilament nylon. Debris attached to the knot includes red blood cells.

KNOT TIED IN SURGICAL THREAD OF 10/0 GAUGE THICKNESS

600X

Drug delivery capsule

SOME DRUGS WORK BEST when delivered to a specific site in the body. They can be encased in a capsule that is designed to burst only at that point, in response to its environment. This capsule can contain not just the drug but also other capsules containing other drugs, which can be targeted at other parts of the body.

BURST DRUG DELIVERY POLYMER CAPSULE REVEALING SMALLER CAPSULES INSIDE IT

195

Surgical clamp

THE CLAMP IS MADE OF A NICKEL-TITANIUM alloy and its sheath (yellow) is made of polytetrafluoroethylene (PTFE) plastic. Springs keep the jaws of the clamp open when it is not in use. The jaws are closed (as seen here) by sliding the plastic sheath over the clamp.

CLAMP USED IN BRAIN MICROSURGERY. IT IS JUST 0.63 MILLIMETERS IN DIAMETER AND IS USED FOR GRABBING SMALL TUMORS

64

Crown wheel of a watch

WHEN THE WATCH IS WOUND, the rotation of the crown wheel causes the ratchet wheel (at top left, behind it) to turn, and this wheel is attached to the barrel which holds the mainspring of the watch. The mechanism is clean; the tiny dust particles would not affect its function. The absence of screwdriver marks in the crown wheel's slot shows that the assembly has never been serviced.

CROWN WHEEL OF A SWISS-MADE, 17-JEWEL INCABLOC WATCH. THE CROWN WHEEL IS PART OF THE WATCH'S WINDING MECHANISM

15X

Watch gears

THESE COGS ARE PRECISION engineered to move the hands of a watch to keep an accurate record of the time. Cog gears are wheels with teeth that mesh together to transfer motion within the watch. The contour of these teeth is smooth and rounded to enable them to interlock (mate) and release with ease. Small gears turn fast and larger gears turn slower, and in doing so give precision to the watch mechanism.

ASSEMBLY OF GEARS IN A WRISTWATCH

20X

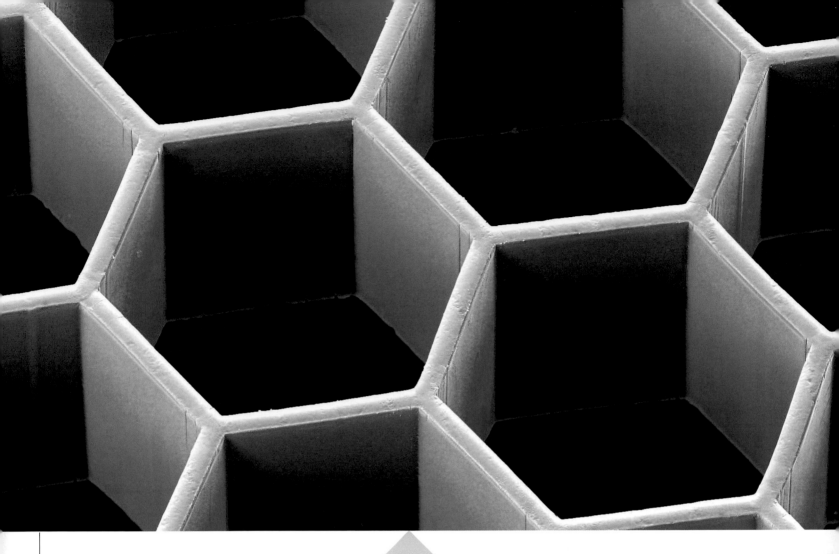

Honeycomb filter

COPIED FROM NATURE, STRUCTURES like this can be used by engineers as particle filters for fuel and as filters for far infrared radiation in astronomy. The hexagonal chambers may also store materials needed for other processes. The honeycomb structure was made using LIGA technology. LIGA is a German acronym for the phrase "Lithographie, Galvanoformung, Abformung." During the LIGA process, X-rays are shone in a pattern onto a substrate. The X-rays alter the properties of the exposed areas so that they can be washed away to leave a microstructure.

THREE-DIMENSIONAL HONEYCOMB STRUCTURE MADE OF COPPER RESEMBLING THE HONEYCOMB OF BEES

350X

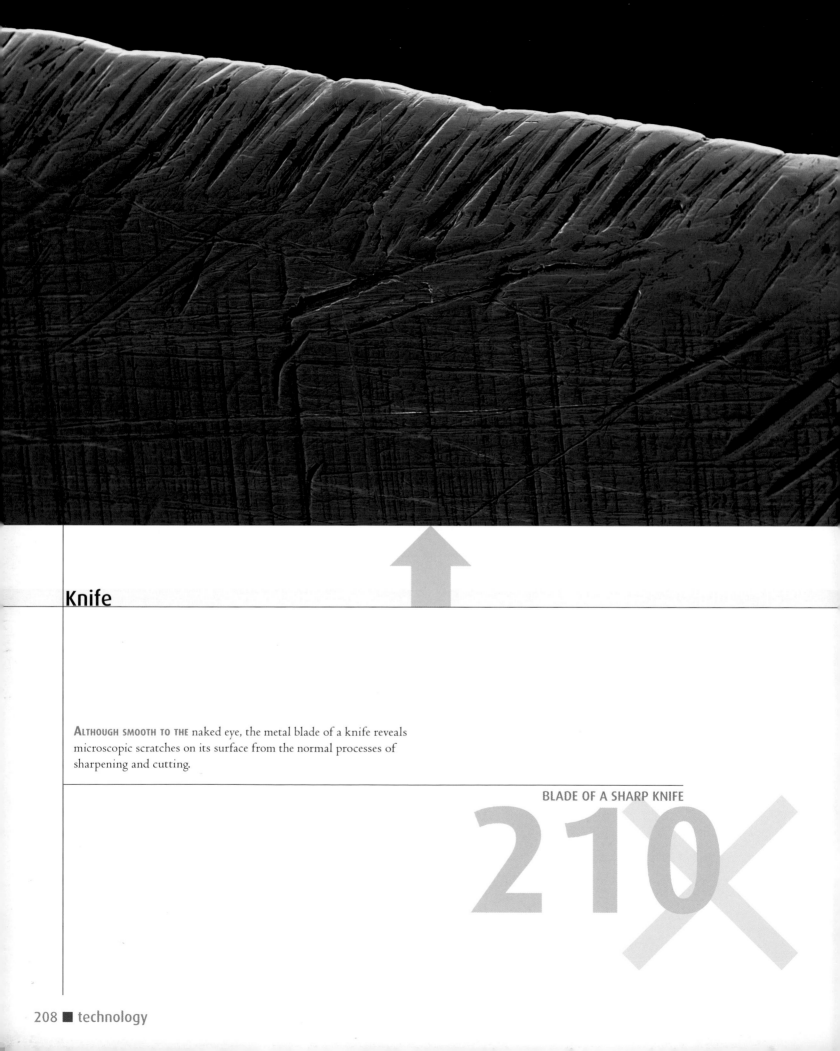

Knife

ALTHOUGH SMOOTH TO THE naked eye, the metal blade of a knife reveals microscopic scratches on its surface from the normal processes of sharpening and cutting.

BLADE OF A SHARP KNIFE

210✕

Used razor

SLICED SHAVINGS OF HAIR, neatly cut at a clean angle, are seen with shaving foam between the multiple blades of a razor.

CUT HAIR ON THE BLADES OF A RAZOR

56

Lightbulb filament

A **LIGHTBULB GLOWS WHEN** the filament inside it is heated to white-hot by the electric current which flows through it. This coiled wire is made of tungsten, a metal element with a high melting point.

COILED FILAMENT OF A LIGHTBULB
480X

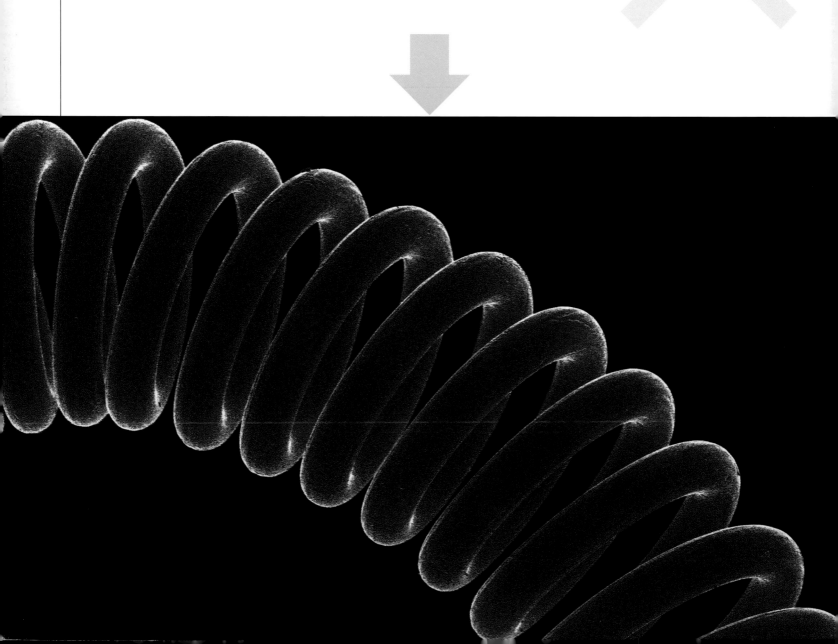

Laboratory microchip

THIS IS USED IN LABORATORY ANALYSIS, and is commonly called a "lab on a chip." Samples and analytical chemicals mix in the chambers and flow through the microchannels. The small size of these channels (typically less than 1 millimeter across) allows many properties to be measured, including fluid viscosity, pH, enzyme reaction kinetics, the molecular diffusion coefficient, number of cells, and protein properties. Manufacturerd by microTEC, Duisburg, Germany.

SURFACE OF A MICROFLUIDIC MICROCHIP, A TYPE OF MICROELECTROMECHANICAL SYSTEM (MEMS)

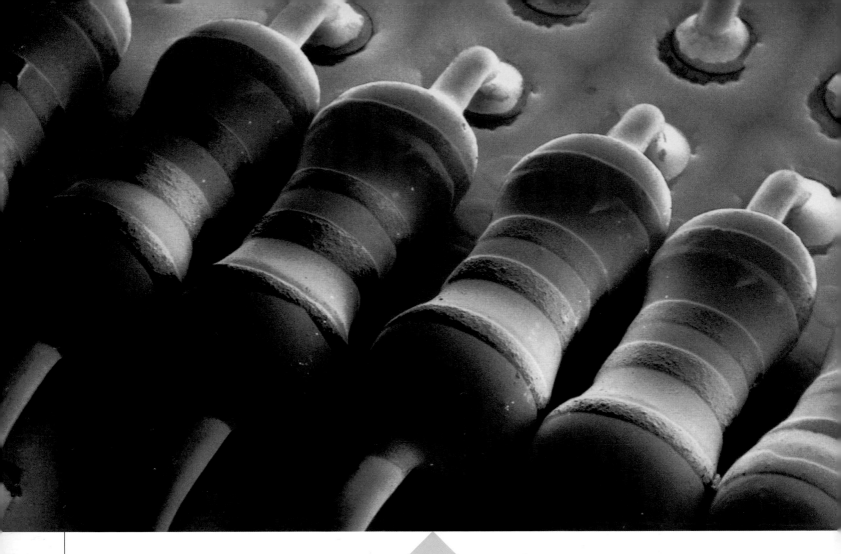

Resistors

Resistors are circuit board components that impede the flow of an electrical current around a circuit. The particular banding pattern on these resistors indicates that their resistance is 22,000 ohms. These are film resistors, consisting of resistive material deposited on an insulator. Circuit boards occur in a huge variety of electronic products, and apart from resistors they may hold microchips.

LINE OF RESISTORS ON AN ELECTRONIC CIRCUIT BOARD

20

Ant holding a microchip

THE SCALE OF THE MINIATURIZATION of microchips (integrated circuits) is well illustrated here, and technological achievements continue to make them smaller. Microchips are used in computers and many other electronic devices, carrying complex microscopic circuits printed onto thin wafers of silicon.

WOOD ANT (FORMICA FUSCA) HOLDING A SILICON MICROCHIP IN ITS JAWS

32

Silicon chip

MICROWIRES IN INTEGRATED CIRCUITS are often made of gold, an excellent conductor of electricity. The wires connect the integrated circuit (chip, at top frame) to pins that allow it to be plugged into a circuit board.

CONNECTING WIRES ON A SILICON COMPUTER MICROCHIP

220✗

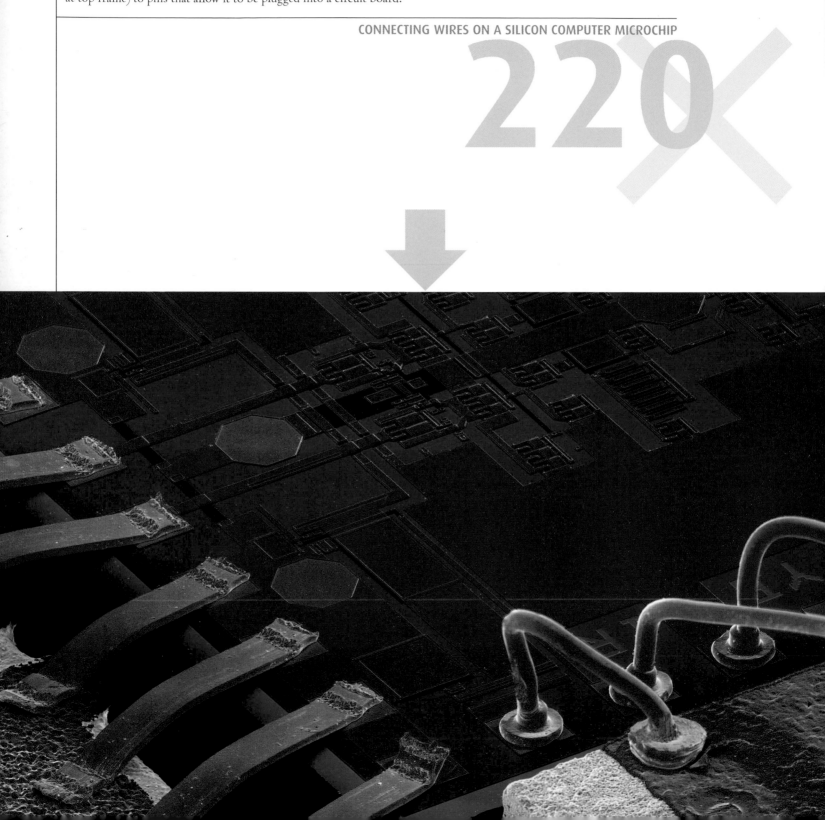

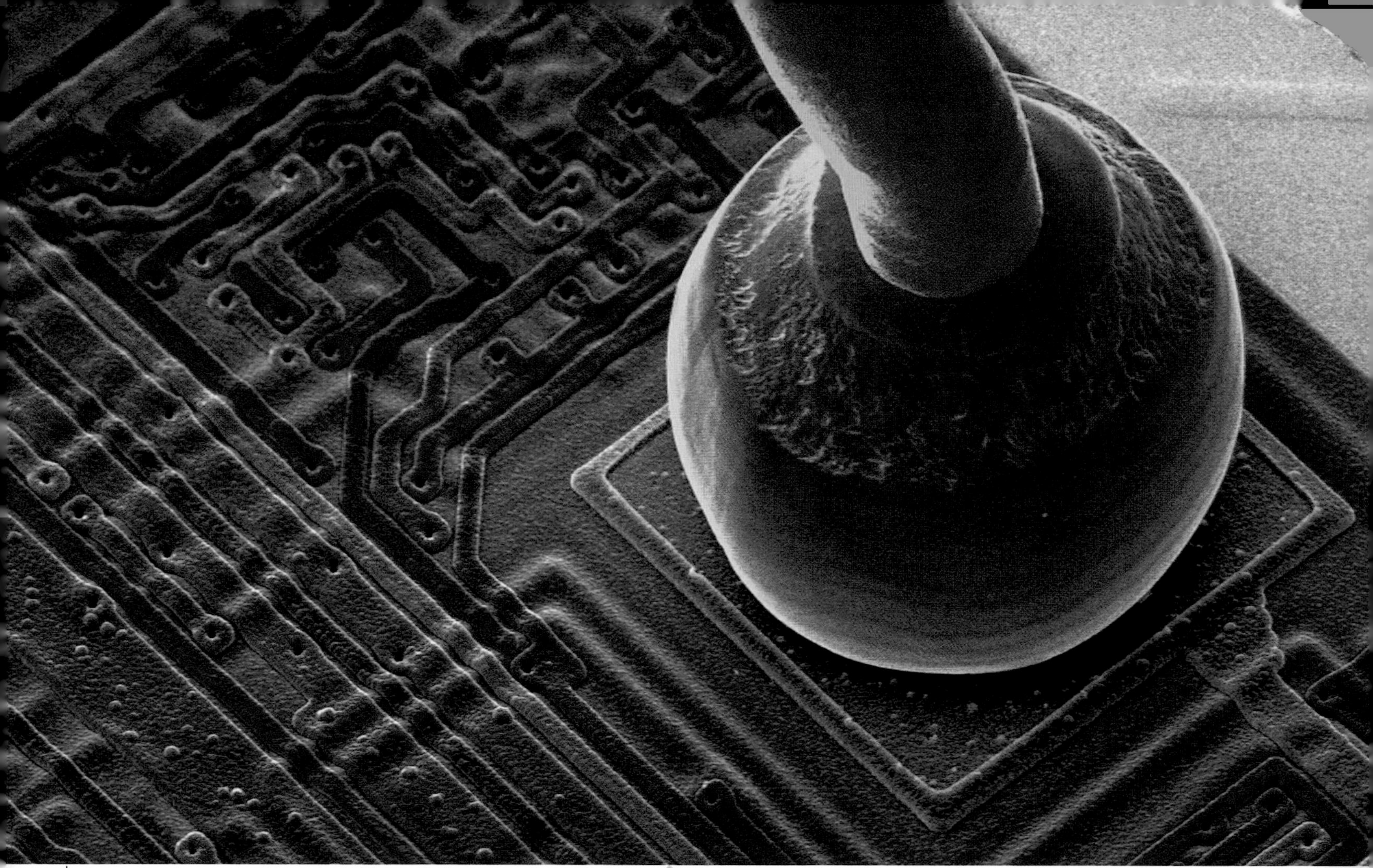

Microwire to microchip

MICROWIRES, OFTEN MADE OF GOLD, connect the integrated circuit (seen here)
to pins (not seen) which allow the chip to be plugged into a larger circuit
board as found in computers. This is integrated circuit M3I6 S32A.
The red, pink, brown and green tracks form the chip's microcircuitry.

BONDED END (GREEN) OF A MICROWIRE (GOLD) CONNECTED TO A SILICON CHIP

835

EPROM silicon chip

THESE COLORED TRACKS OR pathways, geometrically arranged, form part of the chip's microcircuitry. The pits in the tracks are connection points, where connections are made to circuit elements on the other side of the substrate (red). Integrated circuits are made by impregnating the silicon wafer with impurities to create transistors. Tens of thousands of transistors can be fitted on a single chip and connected by conducting pathways.

SURFACE OF AN EPROM (ERASABLE PROGRAMMABLE READ-ONLY MEMORY) SILICON MICROCHIP

4500X

Record stylus

T**HE GROOVES IN THE POLYVINYL CHLORIDE** (PVC) plastic of the record vary in shape, and this is detected as vibrations by the stylus and transformed to produce a sound output. The straighter the groove the quieter the music, the wavier the groove the louder the music. The original groove was produced by reversing the process to record a sound input. In this way, a piece of music or song can be recorded as variations in a groove marked on a plastic disc. This record can then be played to reproduce the original sound.

NEEDLE (STYLUS) OF A STEREO RECORD PLAYER IN A GROOVE ON A VINYL RECORD

225

Compact disc

COLORS ON THE SURFACE are produced by diffraction interference between reflected rays of light from "grooves" in the disc. Like a vinyl record, a compact disc is made of plastic, but the mechanisms of etching the music onto the surface and reading it are quite different. Unlike a vinyl record which has surface grooves read by a stylus (needle), tiny depressions are etched deeper into a compact disc in rows ("grooves") and the digitized signal is read by a laser. Because a CD is an optical disc holding digital data, music, text and images can be stored.

LIGHT MICROGRAPH OF THE SURFACE OF AN AUDIO COMPACT DISC SHOWING MUSIC GROOVES (CENTER) AND SOUNDLESS GROOVES (GREEN BANDS, AT RIGHT). THE RED PAINT IS WRITTEN TRACK INFORMATION

65X

Cracked compact disc

A DEFECT, RECTANGULAR IN SHAPE, is seen in the outer transparent layer of plastic of the compact disc which has peeled away from the surface. Beneath are a series of fine depressions (notches) pressed into a deeper region of the disc which represents a digitized musical signal capable of being read by laser. To reflect the laser light, the music layer is coated with a fine film of metal which follows the depressions exactly. The music is thus sandwiched between two layers of plastic, which prevents dust and scratches from affecting the sound.

SURFACE OF A COMPACT DISC CRACKED TO SHOW THE MUSICAL LAYER BELOW (CENTER)

Microsubsaltwater in an artery

THIS TINY SUBSALTWATER WAS MADE by computer-guided lasers. Laser light caused an acrylic liquid to polymerise, building up the subsaltwater in layers 10 micrometers thick. Tiny nanorobots such as this could be used for detecting and repairing defects in the human body. Powered by a small propeller, they could travel to sites of blockage or damage in blood vessels and repair them from within, restoring correct blood flow. This subsaltwater was made by microTEC of Duisburg, Germany.

CONCEPTUAL IMAGE OF A MICROSUBSALTWATER NANOROBOT IN A HUMAN ARTERY

Nanowires

THE WIRES ARE MADE from rare earth silicides, a compound containing a rare earth metal (lanthanide) combined with silicon. They were laid down on a silicon surface. To create this ultra-high magnification image using a scanning tunneling microscope, the shape of a surfacedown to the level of atoms is mapped by maintaining the flow of current through a fine wire passed just above the surface. These nanowires were made by Hewlett-Packard, Palo Alto, California, USA.

NANOWIRES JUST 10 ATOMS WIDE. THESE WIRES COULD BE USED IN COMPUTERS OPERATING AT THE LIMITS OF MINIATURIZATION

724000X

Carbon nanotubes

THESE TUBES COMPRISE ROLLED SHEETS of carbon atoms structurally related to fullerenes. As cylindrical carbon molecules they have properties which make them potentially very useful in a variety of nano- (microscopic) applications, such as in nano-electronics, enabling them to be used in far smaller components than currently available. They have extraordinary strength, unique electrical properties, and conduct heat efficiently. The name nanotube is derived from their size since their diameter is on the order of a few nanometers (about 50,000 times smaller than the width of a human hair).

SCANNING TUNNELING MICROGRAPH OF CARBON NANOTUBES (RUNNING DIAGONALLY TOWARD LOWER RIGHT). INDIVIDUAL ATOMS ARE SEEN AS BUMPS ON THE SURFACE OF THE TUBES

over 22.5x million

PHOTO CREDITS

All the illustrations of this volume are supplied by **The Science Photo Library** of London, England and are from the following photographers:

Paul Andrews, University of Dundee

Christian Bardele

Dee Breger

Jeremy Burgess

CDC, C. Goldsmith, J. Katz,S. Zaki

Clouds Hill Imaging Ltd.

CMEABG - UCBL1, ISM

CNRI

Michael W. Davidson

John Durham

Eye of Science

Gary Gaugler

Steve Gschmeissner

Hewlett-Packard Laboratories

Manfred Kage

Kenneth Libbrecht

David McCarthy

P. Motta, University 'La Sapienza', Rome

Sidney Moulds

MPI Biochemistry, Volker Steger

Gopal Murti

NIBSC

Susumu Nishinaga

Francois Paquet-Durand

Alfred Pasieka

Steve Patterson

David Scharf

Science Photo Library

Sinclair Stammers

Miodrag Stojkovic

Andrew Syred

Keith Wheeler

Torsten Wittmann

Mosquito head pg.148